Homework Helpers: Basic Math and Pre-Algebra

HOMEWORK HELPERS

Basic Math
and
Pre-Algebra

By

Denise Szecsei

CAREER
PRESS

Pompton Plains, NJ

HOMEWORK HELPERS: BASIC MATH AND PRE-ALGEBRA
TYPESET BY EILEEN MUNSON
Printed in the U.S.A.

To order this title, please call toll-free 1-800-CAREER-1 (NJ and Canada: 201-848-0310) to order using VISA or MasterCard, or for further information on books from Career Press.

P CAREER
PRESS
The Career Press
220 West Parkway, Unit 12
Pompton Plains, NJ 07444
www.careerpress.com

Library of Congress Cataloging-in-Publication Data
Szecsei, Denise.
    Homework helpers. Basic math and algebra / by Denise Szecsei
                p. cm.
        Includes index.
ISBN 978-1-60163-168-8 -- ISBN 978-1-60163-658-4 (ebook) 1. Mathematics. 2. Mathematics--Problems, exercises, etc. I. Title. II. Title: Basic math and algebra

QA39.3.S94 2011
510--dc22

                                                                2010054620

Acknowledgments

Writing a book is a labor-intensive process, not only for the author, but for everyone else involved, directly and indirectly, with the project. This book would not have been completed without the help of a special group of people.

I would like to thank Michael Pye, Kristen Parkes, and everyone else at Career Press who helped transform these mathematical ideas into the object you are now holding. I am most appreciative of the efforts of Jessica Faust, who coordinated things throughout the process.

I am fortunate to be part of the faculty at Stetson University. It is because of the encouragement of the University and my department that I can take the time to write about my favorite subject and hopefully make it one of yours. Stetson University's commitment to education is manifested by the support given to me throughout this process.

Kendelyn Michaels played a pivotal role throughout this project. Because of her thorough reading of the manuscript, I was able to clarify some sections and improve on my explanations. This book would not have been the same without her influence.

Alic Szecsei was in charge of finding all of the typographical errors in the manuscript and checking over my solutions. Writing this book was actually a ruse to get him to solve more math problems.

Thanks to my family for their help throughout the writing and editing stages. The deadline for this manuscript was just before the holidays, and their presents will reflect the time and effort they put into helping me complete this project.

Contents

CONTENTS

C
O
N
T
E
N
T
S

CONTENTS

Welcome to Homework Helpers Basic Math and Pre-Algebra!

Math. One word that, when spoken, can invoke extreme reactions of dread and terror. Most people would rather fight an alien invasion than work out a math problem. In this day and age, working out math problems is more common than fighting alien invasions, so, although it is advantageous to be prepared for an alien attack, you will derive more benefit from developing your math skills.

When children first start learning about numbers, it is fun. Counting things, such as the number of candles on a birthday cake or the number of presents under the tree, is exciting. There is a feeling of accomplishment in being able to count higher and faster than anyone else. But then, something changes.

Math stops being a game and starts to become serious. Getting the right answer becomes important, especially when it involves receiving the correct change from a twenty. Balancing a checkbook, calculating the finance charges on a credit card balance, figuring out the price of an item on sale, and budgeting your monthly expenses can take all of the fun out of working with numbers, but you can't let that stop you from developing the skills necessary to function in our number-centered world.

In this book, I try to hit on the major ideas behind working with numbers, but I start with basic calculations. After you are comfortable adding, subtracting, multiplying, and dividing numbers, you will be ready to solve more interesting problems. As you gain your confidence, I will touch on some of the topics that will prepare you to move on to the study of algebra. Algebra marks the transition from learning how to do basic calculations to being able to solve more complex problems. I'm sure that the glimpse of algebra that you see in this book will get you exited about the prospect of learning more math.

Pre-algebra involves taking a step back from doing rote calculations and looking at the processes involved in those calculations. Mathematics is all about discovering patterns and pushing the limits to develop new methods for solving problems, and pre-algebra will give you a taste of more advanced problem-solving.

In this book, I also cover a little geometry. Geometry and pre-algebra go hand in hand, especially when trying to calculate answers to questions such as:

➲ How many sheets of plywood will you need to build a dog house?

➲ How many bundles of shingles will you need to re-roof your house?

➲ How much money will it cost to tile your dining room?

I don't have a particularly problematic life, yet I find that I use the techniques discussed in this book to solve problems on a daily basis. And that's not just because I teach math! I enjoy using math to solve problems, and I usually turn situations where I use math into word problems to talk about in my classes. Word problems provide you (the reader) with a glimpse of how math can be used to solve problems that occur in everyday life. They also provide a creative outlet for math teachers.

I wrote this book with the hope that it will help anyone who is struggling to understand calculations using basic math or needs to have his or her math skills refreshed. Reading a math book can be a challenge, but I tried to use everyday language to explain the concepts being discussed. Looking at solutions to algebra problems can some-times be confusing, so I tried to explain each of the steps I used from start to finish. Keep in mind that learning math is not a spectator sport. In this book, I have worked out many examples, and I have supplied practice problems at the end of most lessons. Work these problems out on your own as they come up, and check your answers against the solutions provided. Aside from any typographical errors on my part, our answers should match.

Perhaps you aren't quite convinced that math isn't something to be feared. All I am asking you to do is give it a chance. Take a look at the first chapter, try some of the practice problems, and see what happens. Mathematics: The more you know, the better it gets!

1

Numbers

Almost every math book starts out by talking about numbers, and this book is no exception. Math and numbers go hand in hand. That is because numbers are symbols used in math in the same way that letters are symbols used in language. What better numbers to start with than the numbers we learn first: the counting numbers?

The **counting numbers** are the collection of numbers 1, 2, 3, 4.... These numbers go by several names. Besides the counting numbers, they can be called the **natural numbers**, or the **positive integers**. No matter what you call them, the collection of numbers remains the same. These are the numbers that we use when we first learn to count. The dots after the number 4 mean that the pattern continues and that these numbers go on and on. You will never run out of counting numbers. You will always run out of things to count before you run out of numbers to count them.

Notice that the counting numbers start with 1. Zero is a special number, but it is not a counting number. The more you develop your math skills, the more you will appreciate the special qualities of 0. Just adding 0 to the collection of counting numbers is significant enough to warrant giving the entire collection of numbers a new name. We call this list the whole numbers. The **whole numbers** are the numbers 0, 1, 2, 3....

Because we can go on counting forever, you can see that it quickly became necessary to organize the numbers. It would be very impractical to always have a new random symbol for each additional object counted. Rules were established, and using only ten symbols, an effective method for communicating amounts was established. The first object is 1, the second object is 2, and so forth, but once 9 is reached the next object counted is 10. Writing 10 combines two symbols to denote one number. In the number 10, the symbol 1 is located in the **tens' place**, and the symbol 0 is located in the **ones' place**. As we continue to count, we increase the number located in the ones' place until we reach 19. The next number is written 20. The number in the tens' place is increased by 1 and we start over with 0 in the ones' place. Our numbers can increase by changing the value in the ones' place. These same ten symbols are recycled in this pattern to describe numbers through 99. At this point, we need another place. One hundred is represented by the number 100: The symbol 1 is located in the hundreds' place, and there are 0s in the tens' place and the ones' place. We continue with this pattern until we get to 999. The cycle continues and the thousands' place is introduced. As the dots signify, this is how the list of counting numbers can go on forever.

Lesson 1-1: Symbols Used

Just as language has other symbols besides letters (there are periods, semicolons, and question marks), math uses other symbols besides numbers. These symbols tell you what to *do* with the numbers. The four most common operations are addition, subtraction, multiplication, and division. Although there is just one symbol for addition and one symbol for subtraction, there are several symbols for multiplication and division. In addition to manipulating numbers, you can also compare them. To do that, you need to know the symbols for "less than" and "greater than." I'll list the main symbols that we will use here, and give some examples of how they are used.

Symbol	Meaning	Example
+	Addition	$7 + 2 = 9$
−	Subtraction	$7 - 2 = 5$
$\times, \cdot, \text{or} ( )( )$	Multiplication	$4 \times 2 = 8$ $4 \cdot 2 = 8$ $(4)(2) = 8$
$/, -, \text{or} \div$	Division	$4/2 = 2$ $\dfrac{4}{2} = 2$ $4 \div 2 = 2$
=	Is equal to	$4 = 4$
$\neq$	Is not equal to	$4 \neq 2$
>	Is greater than	$4 > 2$
$\geq$	Is greater than or equal to	$4 \geq 4, 4 \geq 2$
<	Is less than	$2 < 4$
$\leq$	Is less than or equal to	$4 \leq 4, 2 \leq 4$

Lesson 1-2: Addition

To get started, we should begin by reviewing the principles of addition. The symbol for addition is +, and the numbers to be added are called addends. The result of the addition is called the sum. The symbol for what the sum is equal to is the equal sign, =. With addition, the order doesn't matter. The sum $3 + 8$ and the sum $8 + 3$ are the same. Because the order that we add things doesn't matter, we say that addition is commutative.

Sometimes addition is written horizontally: $6 + 3 = 9$. Other times addition is written vertically. When written vertically, the horizontal line represents the equal sign. The vertical representation is particularly useful when you are adding two numbers with two or more digits. It helps keep track of the places for the ones, tens, hundreds, etc. by writing the two numbers in a column format with those places aligned:

$$\begin{array}{r} 6 \\ + 3 \\ \hline 9 \end{array}$$

Example 1

Find the sum: $23 + 35$

Solution: $\begin{array}{r} 23 \\ +35 \\ \hline 58 \end{array}$

If the sum of the digits in the ones' column is greater than 9, you will need to carry over to the tens' column. To carry over is to bump up the number in the place immediately to the left of the column under consideration. Carry-over from the ones' place will increase the number in the tens' place. Carry-over from the tens' place will bump up the number in the hundreds' place, and so on. The next examples will help illustrate this point.

Example 2

Find the sum: $54 + 29$

Solution: When we add the numbers in the ones' place, we have $4 + 9 = 13$; this means we need to carry a 1 over to the tens' place:

$$\begin{array}{r} \overset{1}{54} \\ + \ 29 \\ \hline 83 \end{array}$$

Example 3

Find the sum: $156 + 68$

Solution: When we add the numbers in the ones' place, we have $6 + 8 = 14$; this means we need to carry a 1 over to the tens' place. We will also need to carry over to the hundreds' place:

$$\begin{array}{r} \overset{1\ 1}{156} \\ + \ 68 \\ \hline 224 \end{array}$$

We can add three numbers together as well. To find the sum $2 + 6 + 4$, we could either first find $2 + 6$ and then add that result to 4, or we could find $6 + 4$ and then add 2 to that result. Either way we would get an answer of 12.

Because we can group addition in any way, we want we say that addition is **associative**.

Lesson 1-2 Review

Find the following sums:

1. $62 + 33$

2. $84 + 18$

3. $136 + 158$

4. $716 + 237$

Lesson 1-3: Subtraction

The symbol for subtraction is –. The number to be subtracted from is called the **minuend**, and the number to be subtracted is called the

NUMBERS

1

subtrahend. The result of the subtraction is called the **difference**. The symbol for what the difference is equal to is the equal sign, =.

Subtraction is the reverse of addition. If $5 + 3 = 8$ then $8 - 3 = 5$ and $8 - 5 = 3$. You can always check your answers by adding the difference and the subtrahend together; their sum should be the minuend.

Even though subtraction is the reverse of addition, there are some differences. With subtraction, *the order does matter*; subtraction is *not* commutative. We also cannot group things arbitrarily; subtraction is not associative. We will talk more about the order in which we calculate in Lesson 1-8.

Sometimes subtraction is written horizontally: $8 - 5 = 3$. Other times subtraction is written vertically:

$$\begin{array}{r} 8 \\ -\ 3 \\ \hline 5 \end{array}$$

The vertical representation is particularly useful when you are subtracting two numbers with two or more digits. When you subtract one two-digit number from another, you can write the two numbers in a column format with the ones, tens, hundreds, and so on aligned. Subtract each column of digits; if the top digit is smaller than the bottom digit, you need to borrow from the column to the left.

Example 1

Find the difference: $68 - 35$

Solution:
$$\begin{array}{r} 68 \\ -\ 35 \\ \hline 33 \end{array}$$

You should make a point of checking your work:
$$\begin{array}{r} 33 \\ +35 \\ \hline 68 \end{array}$$

Example 2

Find the difference: 54 – 29

Solution: When subtracting the numbers in the ones' place, we have
4 – 9. Because 4 is less than 9, we will need to borrow a 10 from the tens'
place and the number in the tens' place will be reduced from 5 to 4:

$$
\begin{array}{r}
\overset{4}{\cancel{5}}\,{}^{1}4 \\
-2\ 9 \\
\hline
2\ 5
\end{array}
$$

Again, check your work:
$$
\begin{array}{r}
\overset{1}{2}5 \\
+29 \\
\hline
54
\end{array}
$$

Example 3

Find the difference: 156 – 68

Solution: When subtracting the numbers in the ones' place, we have
6 – 8. Because 6 is less than 8, we will need to borrow a 10 from the
tens' place and the number in the tens' place will be reduced from 5
to 4. Looking ahead to the subtraction in the tens' place, we will have
4 – 6. Because 4 is less than 6, we will also need to borrow a 100 from
the hundreds' place:

$$
\begin{array}{r}
{}^{0}\cancel{1}\ {}^{1}\overset{4}{\cancel{5}}\ {}^{1}6 \\
-\ \ 6\ \ 8 \\
\hline
8\ \ 8
\end{array}
$$

Check your work:
$$
\begin{array}{r}
\overset{1}{8}8 \\
+\ 68 \\
\hline
156
\end{array}
$$

Lesson 1-3 Review

Find the following differences.

1. $9 - 2$

2. $72 - 31$

3. $63 - 39$

4. $322 - 78$

Lesson 1-4: Multiplication

Multiplication is a shortcut for addition. The expression 4×5 can be interpreted as either $5 + 5 + 5 + 5$ or $4 + 4 + 4 + 4 + 4$. In other words, 4×5 can be thought of as adding four fives or as adding five fours. Being familiar with the multiplication table (or times table) for the first 10 whole numbers will be very helpful in almost every math class you take.

There are three popular symbols used to represent multiplication: $\times$, $\cdot$, and ( )( ). The expressions 4×5, $4 \cdot 5$, and $(4)(5)$ all mean the same thing. The numbers that are being multiplied together are called **factors**, and the result is called the **product**. With multiplication the order doesn't matter; evaluating 4×5 gives the same result as evaluating 5×4. We say that multiplication is commutative (just as addition is).

Multiplying by 10, 100, 1,000, etc. is probably the nicest multiplication of all. Notice that $6 \times 10 = 60$, $6 \times 100 = 600$, and $6 \times 1,000 = 6,000$. There is a pattern worth noting: When multiplying a number by 10 you just take that number and put a 0 at the end. When multiplying a number by 100, you just take that number and put two 0s at the end. When multiplying a number by 1,000, you just take that number and put three 0s at the end.

Sometimes multiplication problems are written horizontally as shown here: $4 \times 5 = 20$. Other times, multiplication is written vertically:

8
× 3
———
24

The vertical representation is useful when a multiplication problem involves one or more two-digit numbers.

Example 1

Find the product: 15×6

Solution: First find the product 5×6, which is 30. Put a 0 in the ones', place of the product and carry over 3 tens. Then, find the product 1×6, which is 6, and add the 3 tens that you carried over to get 9:

$\overset{3}{1}5$
× 6
———
90

Example 2

Find the product: 26×4

Solution: First, find the product 6×4, which is 24. Put a 4 down in the ones' place and carry over the 2 tens. Then, find the product 2×4, which is 8, and add the 2 tens that you carried over to get 10:

$\overset{2}{2}6$
× 4
———
104

To multiply two two-digit numbers together, break the problem up into two problems that involve multiplying one two-digit number and one one-digit number, and add the two results together. There's just one twist: When you work out the second product, you must shift your numbers to the left by putting a 0 in the ones' place. This is because the second number is in the tens' place; you've already taken care of the ones' place with the first product. This method will become clearer once I discuss the distributive property.

Example 3

Find the product: 32 × 43

Solution:

Work out the problem 32 × 3 (ones):	Then, focus on the 4 in 43 and multiply 32 × 4 (tens). Remember to put a 0 in the ones' place in the product:	Now add the two products together:
32 × 43 ——— 96	32 × 43 ——— 96 1,280	32 × 43 ——— 96 +1,280 ——— 1,376

So 32 × 43 = 1,376.

Example 4

Find the product: 68 × 23

Solution: Break the problem up into two products: 68 × 3 (ones) and 68 × 2 (tens) then add the results:

$$
\begin{array}{r}
\overset{2}{68} \\
\times\ 23 \\
\hline
204
\end{array}
\qquad
\begin{array}{r}
\overset{1}{68} \\
\times\ 23 \\
\hline
204 \\
1,360
\end{array}
\qquad
\begin{array}{r}
68 \\
\times\ 23 \\
\hline
204 \\
+1,360 \\
\hline
1,564
\end{array}
$$

So 68 × 23 = 1,564.

When you multiply 3 numbers together, the order in which you perform the multiplication doesn't matter. For example, when finding the product $4 \times 2 \times 5$, it doesn't matter if you first find 4×2 and then multiply the result by $5 \cdot (4 \times 2) \times 5 = 8 \times 5 = 40$ or if you first find 2×5 and then multiply the result by $4 \cdot 4 \times (2 \times 5) = 4 \times 10 = 40$. Multiplication is said to be associative (again, just as addition is). It doesn't matter how you group a string of products together, or which products you find first.

Lesson 1-4 Review

Find the following products.

1. 4×9

2. 36×8

3. 62×18

4. $3 \times 5 \times 8$

5. $16 \times 10,000$

Lesson 1-5: Properties of 0 and 1

You are now ready to learn some of the properties of zero that make it unique. The symbol zero represents that there is "nothing," but it is very important to know *where* we have nothing. You use 0 as a placeholder in a number such as 1,052. If you didn't have a zero in the hundreds place, the number would be 152. You have just seen examples of how zero is used as a placeholder in multiplication. The thousands', hundreds', tens', and ones' places are a way to organize our numbers so that we can perform calculations more easily. Without 0, this system could not work.

In addition to being a placeholder, the "nothingness" of 0 is also special because it can be added to any number and not change the original number. For example, $6 + 0 = 6$ and $152 + 0 = 152$. Zero is called the **additive identity** because when you add it to any number you don't change that number. In our number system, there is only one additive identity; 0 is unique in that regard.

Zero does some strange things when it is involved in multiplication. Instead of doing "nothing" to the original number, it turns all numbers into itself. If you take any number and multiply it by 0, you will get 0. Zero is the great leveler of multiplication. You can take a number as big as three trillion, but simply multiplying it by 0 leaves you with 0. I can't say it enough: 0 times any number is 0. So, adding 0 to a number doesn't change it, but multiplying a number by 0 is 0.

I have given you plenty of reasons why 0 is a pretty special number, but there's even more to admire about it. If you multiply any two numbers together and get 0, then you know that one of the numbers involved in the multiplication must have been 0. In other words, the only way that the product of two numbers is 0 is if one of the two numbers in the product is 0. Wow—0 *is* a many splendid thing.

Zero has gotten all of the attention in this lesson, but there is another number mentioned in the title: 1. The number 1 can also do some amazing things. Multiplying a number by 1 doesn't change the number: $6 \times 1 = 6$ and $30 \times 1 = 30$. The number 1 is the only number with this particular property. We call 1 the **multiplicative identity** because nothing changes when you *multiply* a number by 1. Zero is to addition as 1 is to multiplication. It will be important to remember these properties because they come in handy in algebra.

Lesson 1-6: Division

Division is the reverse, or undoing, of multiplication. Division is related to multiplication in the same way that subtraction is related to addition. Because $4 \times 5 = 20$ we say that $\frac{20}{5} = 4$ and $\frac{20}{4} = 5$. Being familiar with the multiplication table will be very helpful when solving division problems.

The symbols used for division are $\div, -, /,$ or $\overline{)}$. The expressions $20 \div 5$, $\frac{20}{5}$, $\frac{20}{5}$, and $5\overline{)20}$ all mean the same thing. The number 20 in all of the examples is the number that gets divided; it is called the **dividend**. The number 5 in all of the examples is the number that divides into the dividend; it is called the **divisor**. The answer obtained after doing the division is called the **quotient**. Just as with subtraction, order matters; division is *not* commutative.

To work out a division problem involving two-digit numbers, it is best to use long division. I will review long division in the first example.

Example 1

Divide: 348 ÷ 5

Solution: Write the problem as $5\overline{)348}$. Divide the divisor (in this case, 5) into the left digit (or digits) of the dividend (in this case, 348). To do this, choose the smallest part of the dividend that the divisor will divide in to. Because 5 is greater than 3, we need to use the first two digits of our dividend: 34. Look for the largest multiple of 5 that is less than 34. Because $5 \times 6 = 30$ and $5 \times 7 = 35$, the largest multiple of 5 that is less than 34 is 30. Place a 6 (the other factor involved in the product that gives 30) above the 4 (the last digit of the dividend under consideration), put 30 underneath the 34 and subtract 30 from 34, and bring down the 8 from the original 348:

```
    6
5)348
   30
   48
```

Now do the same thing with the next two digits. Look for the largest multiple of 5 that is less than 48. Because $5 \times 9 = 45$, that's what we need. Write a 9 next to the 6 and 45 underneath the 48, and subtract:

```
    69
5)348
   30
   48
   45
    3
```

Because 3 is less than 5, and there are no more digits to the right to incorporate, we are done. The number 3 in this example is called the remainder. The **remainder** is the number "left over" when we are finished doing the long division. So 348 ÷ 5 is equal to 69 remainder 3.

A smaller number is a **factor** of a larger number if, when you divide the smaller number into the larger number you end up with a remainder of 0. It is called a factor because you can multiply it by the quotient to get the larger number. Remember that multiplication and division go hand-in-hand. For example, 4 is a factor of 20 because 20 ÷ 4 equals 5 remainder 0. Looking at it another way, 4 × 5 = 20. We say that 20 is *evenly divisible* by 4. Every number is evenly divisible by its *factors*. "Divides evenly" means the same thing as "has remainder 0." Every whole number is evenly divisible by 1, and every non-zero whole number is evenly divisible by itself.

Lesson 1-7: Prime Numbers and Factoring

A number is **prime** if the only numbers that divide into it evenly are 1 and itself. In other words, a prime number has no other factors besides 1 and itself. For example, 3 is prime because the only numbers that divide it evenly are 1 and 3.

> Remember that every whole number, other than 0 and 1, has at least two factors: 1 and itself.

Because the number 1 divides evenly into every other number, we call 1 a **trivial factor**.

The number 1 is in a class by itself. There is only one number that evenly divides 1: itself. It is not considered to be prime.

If a number other than 1 is not prime, it is **composite**. A composite number is a number that can be evenly divided by a number other than 1 or itself. For example, 9 is composite because, in addition to 1 and 9, it is evenly divisible by 3.

We can list the first few prime numbers: 2, 3, 5, 7, 11, 13, 17, 19, 23.... The list of prime numbers never ends. Over the years, many mathematicians have been, and continue to be, fascinated with prime numbers, and new discoveries about prime numbers are still being made. The largest prime number known requires more than 4 million digits to write out completely, and mathematicians are currently looking for an even larger prime number.

Prime numbers are useful for factoring numbers. Every composite number can be written as a product of prime numbers. For example, $18 = 2 \times 3 \times 3$, and 2 and 3 are prime numbers. In order to factor a composite number, you need to go down the list of prime numbers, divide each prime number into the composite number, and check to see if the remainder is zero. If the remainder is 0, the prime number you used is a factor of the composite number.

There is a systematic approach for finding factors.

⊃ First, check to see whether the composite number is even or odd. An **even** number is a number that is evenly divisible by 2. Examples of even numbers are 2, 4, 6, 8, 10, 12.... Even numbers are easy to recognize because their last digit is either 0, 2, 4, 6, or 8. Numbers that are not even are called **odd** numbers. Odd numbers are also easy to recognize: Their last digit is either 1, 3, 5, 7, or 9. If a composite number is even, it will be divisible by 2 (the first prime number).

⊃ If a composite number is not even, check to see if it is evenly divisible by 3. One way to do this is to actually do the division. If the remainder is 0, then 3 is a factor of the number. If the remainder is not 0, then 3 is not a factor. There is a shortcut available to see if 3 is a factor of a number: Add the digits of the number together. If the resulting sum is evenly divisible by 3, then so was the original number. For example, to check whether 344 is divisible by 3, add the digits together: 3 + 4 + 4 = 11. Because the remainder of 11 ÷ 3 is 2 (not 0), 3 does not evenly divide 344. To check whether 357 is divisible by 3, add the digits together: 3 + 5 + 7 = 15. Because 15 ÷ 3 has remainder 0, 3 divides 357. We can use long division to be sure:

$$\begin{array}{r} 119 \\ 3\overline{)357} \\ \underline{3} \\ 05 \\ \underline{3} \\ 27 \\ \underline{27} \\ 0 \end{array}$$

⊃ Once you are done with 2 and 3, move on to 5. Check to see whether the composite number is evenly divisible by 5. Again, there is an easy check: If the number ends in either 0 or 5, it will be divisible by 5. Then move on to the next prime number, 7. Continue down the list until you have checked all of the prime numbers smaller than the composite number. If you can't find a prime number that divides your number evenly, then your number is not a composite number.

If, on the other hand, you successfully find a prime number that divides into the composite number, do the division to find the other factor that, when multiplied by the prime number you just discovered, gives your original number. Start to factor this new number. Keep going until the last factor you are left with is a prime number. This process is called **creating a factor tree**. When you are done, you will have a list of prime numbers that multiply to give you your original number.

Example 1

Create a factor tree for 420.

Solution: 420 is even because it ends in 0, so let's start by dividing it
by 2:

$$
\begin{array}{r}
210 \\
2\overline{)420} \\
4 \\
\hline
02 \\
2 \\
\hline
00 \\
0 \\
\hline
0
\end{array}
$$

So $420 = 2 \times 210$. Now we need to factor 210. Because 210 is even, it
is also evenly divisible by 2:

$$
\begin{array}{r}
105 \\
2\overline{)210} \\
2 \\
\hline
010 \\
10 \\
\hline
0
\end{array}
$$

Now our factor tree is growing: $420 = 2 \times 210 = 2 \times 2 \times 105$. Now
we need to factor 105. It is not even so it is not divisible by 2. The
sum of the digits is $1 + 0 + 5 = 6$, which is evenly divisible by 3:

$$
\begin{array}{r}
35 \\
3\overline{)105} \\
9 \\
\hline
15 \\
15 \\
\hline
0
\end{array}
$$

Our tree is getting longer:

$420 = 2 \times 210 = 2 \times 2 \times 105 = 2 \times 2 \times 3 \times 35$.

Now we need to factor 35: $5 \times 7 = 35$ and both 5 and 7 are prime numbers, so we are done:

$420 = 2 \times 210 = 2 \times 2 \times 105 = 2 \times 2 \times 3 \times 35 = 2 \times 2 \times 3 \times 5 \times 7$.

The factor tree can be visualized as a tree with branches, as shown in Figure 1.1.

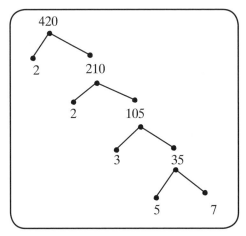

Figure 1.1

You will be surprised at how useful factoring will be.

If two numbers have no common factor other than 1, they are said to be **relatively prime**. For example, 15 and 8 are relatively prime, because the factors of 15 are 3, 5, and 15, and the factors of 8 are 2, 4, and 8. Notice that both 15 and 8 are composite numbers, yet they are relatively prime. The numbers 7 and 21 are not relatively prime, because 7 is a factor of both 7 and 21. So even though 7 is a prime number, it is not relatively prime to 21.

If two numbers are relatively prime, then their greatest common factor is 1. If two numbers are not relatively prime, then their **greatest common factor** is the largest number that evenly divides both numbers. The easiest way to find the greatest common factor is to factor each number into their prime factors. Then match up, prime by prime, the prime numbers that appear in both factorizations. If any prime numbers repeat, you must take that into consideration. For example, the greatest common factor of 12 and 15 is 3, because $12 = 2 \times 2 \times 3$ and $15 = 3 \times 5$; 3 is the largest number that evenly divides 12 and 15. The greatest common factor of 20 and 36 is 4: $20 = 2 \times 2 \times 5$ and $36 = 2 \times 2 \times 3 \times 3$. Both 20 and 36 have a repeating factor of 2, which is why their greatest common factor is 4.

Now that you understand the idea of the greatest common factor of two numbers, it is time to talk about the least common multiple of two numbers. The **least common multiple** of two numbers is the smallest number that they both divide into evenly. The easiest way to find the least common multiple of two numbers is to create a factor tree for both numbers and find the greatest common factor. Then multiply the original two numbers together and divide by the greatest common factor. This result is the least common multiple of the two numbers.

Example 2

Find the least common multiple of 12 and 15.

Solution: As previously discussed, the greatest common factor of 12 and 15 is 3. The least common multiple of 12 and 15 is found by evaluating $(12 \times 15) \div 3 \cdot (12 \times 15) \div 3 = 180 \div 3 = 60$. So the smallest number that both 12 and 15 divide evenly into is 60.

Lesson 1-7 Review

1. Create factor trees for the following numbers:

    a. 96

    b. 215

    c. 310

2. Find the greatest common factor and the least common multiple of 15 and 35.

Lesson 1-8: Order of Operations

We have four basic operations: addition, subtraction, multiplication, and division. If several numbers are involved in any combination of these operations, we need to know which operation to perform first. For example, if we want to evaluate $4 + 3 \times 2$ we need to decide whether we will add 4 and 3 together first and then multiply by 2, or multiply 3 and 2 together first and then add 4. You may want to do it one way, and I may want to do it the other way. We would both get different answers and then argue about who was right. In order to avoid the confusion, an order of operations was established. Using the accepted order of operations, everyone will get the same answer when working out calculations that involve more than one operation.

The accepted order of operations stipulates that multiplication and division, read left to right, take precedence over addition and subtraction. In our example $4 + 3 \times 2$, the multiplication of 3 and 2 must be done before any addition takes place. If you specifically wanted the addition to take place first, you would need to use parentheses: $(4 + 3) \times 2$. The order of operations includes a clause for parentheses: Anything in parentheses must be done first. The hierarchy of operations is:

⊃ Parentheses.

⊃ Multiplication and division, read left to right.

⊃ Addition and subtraction, read left to right.

Just to make sure we're on the same page, let's work out a calculation.

Example 1

Evaluate: $2 + 16 \div 2 \times 10$

Solution: Follow the order of operations:

$$2 + 16 \div 2 \times 10$$

Divide 16 by 2.	$2 + 8 \times 10$
Multiply 8 by 10.	$2 + 80$
Add 2 and 80.	82

Lesson 1-8 Review

Evaluate the following.

1. $20 + 3 \times (4 + 5)$
2. $90 - 4 \times 3 + 2$
3. $(2 + 3) \times 5 - 10 \times 2$
4. $4 \div 2 \times 9$

Lesson 1-9: Variables and the Distributive Property

We are starting to talk about the properties of numbers in general. Some numbers are prime; others are composite. Some numbers are even; others are odd. Talking about numbers in general can be awkward. We need a symbol to represent *any* number. The symbols we use to write specific numbers, such as 5, cannot be used to represent *any* number. A symbol that is used to represent any number is called a **variable**. We can use anything other than a number as a variable. We could use letters at the beginning of the alphabet, such as *a*, *b*, or *c*, to serve as a variable. We could use letters at the end of the alphabet, such as *x*, *y*, or *z*, to serve as a variable. We could use a happy face or a star to serve as a variable. The important thing is that the symbol we use for a variable cannot be any of the ten symbols that we use to write specific numbers.

Let me show you one way a variable can be used. If I say "Let *a* represent any counting number," then *a* could be any number from our list (1, 2, 3, 4...). If I wanted to describe the multiplicative property of 0 (remember that special property: 0 times any number is 0), I use the symbol a to represent any number and write $0 \times a = 0$.

We can also use variables to make general statements about numbers. We invented multiplication as a shorthand notation for addition. We noticed that $3 + 3 + 3 + 3 + 3$ can be shortened to 5×3. This shorthand notation is not specific to the number 3. Any number added to itself 4 times can be written as 4 times that number. This is awkward to express using words, but it is easy to express using variables: If a is any number, then $a + a + a + a = 4 \times a$. The multiplication symbol is usually omitted, and we write $4a$ instead of $4 \times a$. Whenever you see a number and a variable next to each other with nothing in between them, you should imagine a multiplication symbol between them. The expression $3a$ means the same thing as "3 times a," or $3 \times a$. The number in front of the variable is called the **coefficient** of the variable, and it represents how many times the variable is added to itself.

Now that we know what $3a$ and $4a$ mean, let's examine what $3a + 4a$ means. We can write out what $3a$ and $4a$ each mean, and go from there: $3a = a + a + a$, and $4a = a + a + a + a$, so $3a + 4a$ means $(a + a + a) + (a + a + a + a)$. This is just a added to itself 7 times, or $7a$. Notice that $3a + 4a = 7a$. Instead of writing out all of the a's in the sum, it is easier just to add the coefficients of the variable together and write this new coefficient in front of the variable. You will see more examples of variables later on in this lesson, and I wanted you to be prepared.

In order to evaluate an expression that involves both multiplication and addition, you need to use the order of operations. Consider the problem $7 \times (5 + 3)$. Using the order of operations, we would add 5 and 3 first because of the parentheses (to get 8), and then multiply 7 and 8 together to get 56. I would get the same result if I added the products 7×5 and 7×3 together: $35 + 21 = 56$. This isn't just a coincidence. It will always happen.

The distributive property allows us to evaluate expressions such as $7 \times (5 + 3)$ in more than one way. The distributive property is stated in general terms using variables. If a, b, and c are any numbers, the distributive property states that:

$$a \times (b + c) = a \times b + a \times c$$

We say that multiplication distributes over addition. The distributive property can be used to make calculations easier, as we will see in the examples. Sometimes it is easier to find the sum of two products using the distributive property than it is to find one product directly.

Example 1

Use the distributive property to calculate: 25×104

Solution:

$25 \times 104 = 25 \times (100 + 4) = 25 \times 100 + 25 \times 4 = 2,500 + 100 = 2,600$

Example 2

Use the distributive property to calculate: 4×508

Solution:

$4 \times 508 = 4 \times (500 + 8) = 4 \times 500 + 4 \times 8 = 2,000 + 32 = 2,032$

The distributive property can be used to understand the process of multiplying two two-digit numbers. For example, to find the product 36×15 apply the distributive property twice:

$$
\begin{aligned}
36 \times 15 \quad &= 36 \times (10 + 5) \\
&= 36 \times 10 + 36 \times 5 \\
&= 360 + [(30 + 6) \times 5] \\
&= 360 + [30 \times 5 + 6 \times 5] \\
&= 360 + [150 + 30] \\
&= 360 + 180 \\
&= 540
\end{aligned}
$$

Now, look at the method for multiplying two two-digit numbers discussed previously. Notice that the intermediate products are

$$
\begin{array}{r}
36 \\
\times\ 15 \\
\hline
180 \\
+360 \\
\hline
540
\end{array}
$$

exactly the same in both methods. The two procedures actually use the same method; the only difference is in how the work is organized.

Lesson 1-9 Review

Use the distributive property to calculate the following.

1. 24 × 12

2. 42 × 13

3. 62 × 13

Answer Key

Lesson 1-2 Review

1. 95

2. 102

3. 294

4. 953

Lesson 1-3 Review

1. 7

2. 41

3. 24

4. 244

Lesson 1-4 Review

1. 36

2. 288

3. 1,116

4. 120

5. 160,000

Lesson 1-7 Review

1. a. $96 = 2 \times 2 \times 2 \times 2 \times 2 \times 3$

96
2 48
2 24
2 12
2 6
2 3

b. $215 = 5 \times 43$

c. $310 = 2 \times 5 \times 31$

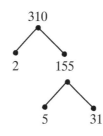

2. Greatest common factor: 5; least common multiple: 105

Lesson 1-8 Review

1. 47

2. 80

3. 5

4. 18

Lesson 1-9 Review

1. 268

2. 546

3. 806

2

Integers

The *whole numbers* only tell you half of the story. The other half of the story involves negative numbers. Most of us are all too familiar with negative numbers. If I have $15 and I want to purchase a pair of shoes that cost $25, then I'm $10 short. We handle the concept of debt using negative numbers. Numbers that are greater than 0 are called positive numbers. The opposite of a positive number is called a negative number. For example, the opposite of 3 is –3, and the opposite of 10 is –10. Opposites work both ways; the opposite of –3 is 3 and the opposite of –10 is 10. A number and its opposite always add up to 0; 3 + (–3) = 0 and 10 + (–10) = 0. The opposite of 0 is 0. It is the only number that is its own opposite—another special property of 0!

The opposite of a number is also called its **additive inverse**. The collection of whole numbers and their additive inverses are called the **integers**. The counting numbers are called the **positive integers** and their opposites are called the **negative integers**. If a number is greater than 0, it has a positive sign; a number that is less than 0 has a negative sign.

Integers have two parts: the magnitude of the integer and the sign of the integer. The **magnitude** of an integer is the whole number component of the integer. It tells how big the number is. The **sign** of the integer indicates whether the integer is positive or negative. The integer 35 has magnitude 35 and sign +. The integer –15 has magnitude 15 and sign –. Most of the time the positive sign is not written and is implied.

Lesson 2-1: The Number Line and Absolute Value

We can graph numbers on a number line. A **number line** is a graphical representation of the signs and magnitudes of numbers and their relationships to each other. You create a number line by first drawing a line and putting a mark for 0, as shown in Figure 2.1. By convention, the positive integers are located to the right of 0; the negative integers are located to the left of 0. The numbers increase in value as you go from left to right, regardless of whether you are working with negative or positive numbers. In general terms using variables, if a number x is located to the left of another number y on the number line, then x is greater than y. From a different perspective, if a number y is located to the right of another number x on the number line, then y is less than x.

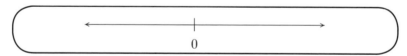

0

Figure 2.1.

To place numbers on a number line, you have to know two things. The first thing you need to know is whether the number goes to the left or the right of 0. This is determined by the sign of the number; if a number is positive it goes to the right, and if it is negative it goes to the left. The second thing you need to know is how far away from 0 to put the number. That is determined by the magnitude of the number. The magnitude of the integer tells you how many units the integer is away from 0.

Example 1

Graph the numbers 2, –3, and 5 on a number line. Order the numbers from smallest to greatest.

Solution:

The numbers from smallest to greatest are –3, 2, 5. Their graph is shown in Figure 2.2.

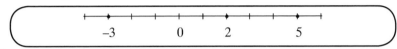

Figure 2.2.

The distance from a number to 0 is called its **absolute value**. The absolute value of a number is *always positive*. A number and its opposite are always the same distance away from 0, but are located on opposite sides of 0. The symbol for absolute value is $|\ \ |$. The absolute value of 3 is written symbolically as $|3|$, and the absolute value of –5 is written symbolically as $|-5|$. Using variables, the absolute value of a number a is written symbolically as $|a|$. The absolute value of a number and its magnitude are the same thing, and the two phrases are used interchangeably.

Example 2

Find: $|4|$ and $|-2|$

Solution: 4 is located 4 units to the right of 0 so $|4| = 4$; –2 is located 2 units to the left of 0, so $|-2| = 2$.

Lesson 2-1 Review

1. Graph the numbers 1, –2, and 4 on a number line. Order the numbers from smallest to greatest.

2. Find $|2|$, $|-6|$, and $|0|$.

Lesson 2-2: Adding Integers

As I mentioned earlier, the sum of a number and its additive inverse, or opposite, is 0; $3 + (-3) = 0$ and $(-4) + 4 = 0$. You can

2 INTEGERS

add two integers together using a number line. To add two integers together, locate the first number on the number line. Next, find the absolute value of the second number. That will tell you how many units to move over. The sign of the second number tells you whether to move to the left or the right. If the second number is positive, move to the right; if the second number is negative, move to the left.

Example 1

Find the sum: 4 + (–3)

Solution: Start at 4 and move 3 units to the left, as shown in Figure 2.3: 4 + (–3) = 1.

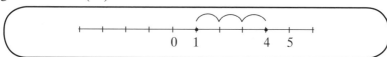

Figure 2.3.

Example 2

Find the sum: (–4) + (–3)

Solution: Start at –4 and move 3 units to the left, as shown in Figure 2.4: (–4) + (–3) = –7.

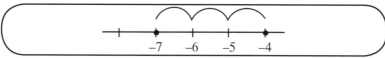

Figure 2.4.

Example 3

Find the sum: (–5) + 3

Solution: Start at –5 and move 3 units to the right, as shown in Figure 2.5: (–5) + 3 = –2.

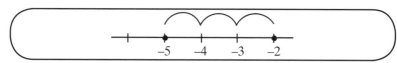

Figure 2.5.

INTEGERS

2

When you use a number line to add two integers together, you are finding a **graphical** solution to the problem. You can also add integers **algebraically**, or *without* using a number line. Using a number line to visualize the process helps keep things straight, but if you are working with large numbers, you may need a really, really long number line.

When adding two integers together there are three different possible scenarios: Both of the integers could be positive, both integers could be negative, or the integers could have opposite signs. You have already practiced adding two positive integers together; adding two positive integers is the same thing as adding two whole numbers. We need to discuss how to add two negative integers together, as well as how to add two integers that have opposite signs.

One way to add two negative integers is to turn it into a problem of adding two positive integers and then changing the sign of the answer. For example, $(-4) + (-3) = -7$ is the same as $-(4 + 3) = -7$. So, when you add two negative integers you can pretend that they are both positive integers, add them together, and then put a negative sign in front.

Example 4

Find the sum: $(-5) + (-6)$

Solution: $(-5) + (-6) = -(5 + 6) = -11$.

When you add two integers that have opposite signs, find the difference between their absolute values. Then, think about the sign of your answer. The integer with the larger absolute value determines whether the answer will be positive or negative. If the integer with the larger absolute value is negative, your answer will be negative. If the integer with the larger absolute value is positive, then your answer will be positive.

Example 5

Find the sum: $18 + (-6)$

Solution: Find the difference between their absolute values: $18 - 6 = 12$. The integer with the larger absolute value is 18, so the answer will be positive. $18 + (-6) = 12$.

Notice that $18 + (-6)$ can be thought of as $18 - 6$. Adding integers with opposite signs and finding the difference between the magnitudes of the two integers gives the same result. You can think of subtraction as adding two integers that have opposite signs.

Example 6

Find the sum: $(-12) + 5$

Solution: Find the difference between their absolute values: $12 - 5 = 7$. The integer with the larger absolute value is -12, so our answer should be negative: $(-12) + 5 = -7$.

Example 7

Find the sum: $3 + (-7)$

Solution: Find the difference between their absolute values: $7 - 3 = 4$. The integer with the larger absolute value is -7, so our answer should be negative: $3 + (-7) = -4$.

You will often see the parentheses left off of these problems. We can write $(-12) + 5$ as $-12 + 5$ and $3 + (-7)$ as $3 - 7$.

Lesson 2-2 Review

Find the sums.

1. $8 + (-3)$
2. $-2 + (-5)$
3. $-3 + (-8)$
4. $-6 + 4$
5. $15 + (-3)$
6. $-10 + 3$

Lesson 2-3: Subtracting Integers

Subtracting two integers can involve finding the opposite of an opposite. We need to examine what the opposite of a negative number is in more detail. When we write –3, we mean the opposite of 3. The negative sign in front of the 3 indicates that it is the opposite of 3. In general, if *a* is any number, then the *opposite of a* is written –a. For example, –3 is the opposite of 3, and the opposite of –3 is written as –(–3). Earlier, we said that 3 and –3 were opposites of each other, meaning that 3 is the opposite of –3, and –3 is the opposite of 3. This means that –(–3) and 3 are two ways to write the opposite of –3; this leads us to the fact that –(–3) = 3. This equation just means that the opposite of –3 is 3. We can also make the signs more explicit: –(–3) = +3. This will come in handy when we subtract negative integers.

Notice that we use the same symbol for subtraction and for a negative number. This is not because we ran out of symbols. We use the same symbol to remind us that the two concepts are related. Subtracting is the same as adding the opposite. When subtracting a negative number, it may help to realize that subtracting a negative is the same as adding. Two wrongs may not make a right, but two negatives definitely make a positive. Keep this in mind as we work the next examples.

Example 1

Find: $6 - (-2)$

Solution: We are subtracting the opposite of 2. As discussed earlier, the opposite of a negative number is a positive number, and our problem actually turns into one involving addition:
$6 - (-2) = 6 + 2 = 8$.

Example 2

Find: $-6 - (-4)$

Solution: We are subtracting the opposite of 4, so our problem turns into one involving the sum of two integers with opposite signs: $-6 - (-4) = -6 + 4 = -2$.

Example 3

Find: $-5 - 3$

Solution: We can write $-5 - 3$ as $-5 + (-3)$, which is very similar to some of the problems we worked in the last section: $-5 + (-3) = -8$.

Lesson 2-3 Review

Find the following differences.

1. $8 - (-3)$

2. $-2 - (-5)$

3. $-3 - (-8)$

4. $-6 - 4$

5. $15 - (-3)$

6. $-10 - 3$

Lesson 2-4: Multiplying Integers

In order to multiply two integers, I need to go back to what multiplication represents. The product 3×4 is a shorthand way of writing $4 + 4 + 4$. From this, we can interpret a product such as $3 \times (-5)$: $3 \times (-5) = (-5) + (-5) + (-5)$. So if we use a number line to figure this out, we would start at -5, move 5 spaces to the left, and then move 5 more spaces to the left. We would move a total of 10 spaces to the left and end at -15. So $3 \times (-5) = -15$. In this section, I will summarize some of the many patterns involved with multiplication.

The product of two whole numbers is a whole number. Because whole numbers are also called positive integers, we can say that the product of two positive integers is a positive integer. We can expand this idea and say that the product of two integers with the same sign will be a positive integer. In other words, 5×7 is a positive integer (as we saw in Chapter 1) and $(-5) \times (-7)$ is also a positive integer. This is an important observation worth writing out again: The product of two negative integers is a positive integer.

As we saw earlier, $3 \times (-5) = (-5) + (-5) + (-5) = -15$. This one example generalizes to the following idea: The product of two integers with opposite signs will be a negative integer.

These steps will walk you through finding the product of two integers:

1. Find the product of their magnitudes, or absolute values.

2. Determine whether the product should be positive or negative.

    ⇨ If the signs of the two integers involved in the product are the *same*, your answer will be a positive integer.

    ⇨ If the signs of the two integers involved in the product are *different*, your answer will be a negative integer.

Example 1

Find the product: $(-5)(-4)$

Solution: Because the signs of the two integers being multiplied are both negative, the answer will be positive: $(-5)(-4) = 20$.

Example 2

Find the product: $(-6)(5)$

Solution: Because the signs of the two integers being multiplied are not the same, the answer will be negative: $(-6)(5) = -30$

To find the product of three integers, use the fact that multiplication is associative and break the problem up into finding two products.

2

INTEGERS

Example 3

Find the product: $(-5)(2)(-3)$

Solution: Use the associative property of multiplication:

$$(-5)(2)(-3)$$

Use the associative property of multiplication.	$[(-5)(2)](-3)$
Evaluate $(-5)(2)$. A negative times a positive will be negative.	$(-10)(-3)$
Find the product $(-10)(-3)$. A negative times a negative will be positive.	30

Lesson 2-4 Review

Find the following products.

1. $3 \times (-6)$

2. $-4 \times (-3)$

3. $8 \times (-4)$

4. $(-10) \times 4 \times (-3)$

Lesson 2-5: Dividing Integers

When dividing one integer by another integer it is best to divide the problem up into two parts: determining the sign of the answer and determining the magnitude of the answer. To determine the magnitude of the answer, pretend that both numbers involved in the division problem are positive and do the division. To determine the sign of the answer, look at the signs of the two integers involved in the problem. Just as with multiplication, if the two signs are the same, the answer will be positive. If the two signs are different, the answer will be negative. It doesn't matter which number (the dividend or the divisor) is negative; as long as the dividend and the divisor have opposite signs, the answer will be negative.

Example 1

Find the quotient: $(-30) \div (-5)$

Solution: First, find the magnitude of the answer: $30 \div 5 = 6$. Next, determine whether your answer should be positive or negative. Because the signs of both integers involved in the problem are the same (they are both negative), the answer should be positive: $(-30) \div (-5) = 6$.

Example 2

Find the quotient: $(-42) \div 6$

Solution: First, find the magnitude of the answer: $42 \div 6 = 7$. Next, determine whether your answer should be positive or negative. Because the signs of both integers involved in the problem are different, the answer should be negative: $(-42) \div 6 = -7$.

Example 3

Find the quotient: $52 \div (-4)$

Solution: First, find the magnitude of the answer:

$$
\begin{array}{r}
13 \\
4{\overline{\smash{\big)}\,52}} \\
\underline{4} \\
12 \\
\underline{12} \\
0
\end{array}
$$

Next, determine whether your answer should be positive or negative. Because the signs of both integers involved in the problem are different, the answer should be negative: $52 \div (-4) = -13$.

2
INTEGERS

Lesson 2-5 Review

Find the following quotients.

1. $-36 \div 3$
2. $-90 \div (-5)$
3. $27 \div (-9)$
4. $-64 \div (-16)$

Lesson 2-6: Integers and the Distributive Property

The nature of the distributive property doesn't change with the introduction of negative integers. It just takes a little practice adjusting to the variety of problems that arise with this new twist. Remember the distributive property:

If a, b, and c are numbers, then $a \times (b + c) = a \times b + a \times c$.

To evaluate $a \times (b + c)$ you will need to evaluate $a \times b$ and $a \times c$ separately and then add them.

Example 1

Evaluate the following:

a. $(-7) \times (5 + 3)$

b. $(-4) \times (6 - 2)$

c. $(-3) \times (-2 - 8)$

d. $(2) \times (3 - 8)$

Solution:

a. $(-7) \times (5 + 3) = (-7)(5) + (-7)(3) = -35 + -21 = -56$

b. $(-4) \times (6 - 2) = (-4)(6) + (-4)(-2) = -24 + 8 = -16$

c. $(-3) \times (-2 - 8) = (-3)(-2) + (-3)(-8) = 6 + 24 = 30$

d. $(2) \times (3 - 8) = (2)(3) + (2)(-8) = 6 + (-16) = -10$

It may seem as though we have evaluated these expressions the hard way. Using the order of operations would have been much simpler, but using this technique has its advantages, as we will see in the next example. Multiplying two two-digit numbers becomes easier to do using the distributive property.

Example 2

Use the distributive property to find the following products:

a. 25×96

b. 4×490

Solution:

a. $25 \times 96 = 25 \times (100 - 4) = 25 \times 100 + 25 \times (-4)$
$= 2,500 - 100 = 2,400$

b. $4 \times 490 = 4 \times (500 - 10) = 4 \times 500 + 4 \times (-10)$
$= 2,000 + (-40) = 1,960$

Lesson 2-6 Review

Use the distributive property to evaluate the following products.

1. 6×48

2. 12×47

3. 37×8

4. 23×59

2

INTEGERS

Answer Key

Lesson 2-1 Review

1. The numbers 1, –2, and 4 on the number line.

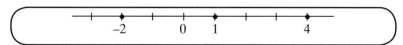

2. $|2| = 2$
   $|-6| = 6$
   $|0| = 0$

Lesson 2-2 Review

1. 5
2. –7
3. –11
4. –2
5. 12
6. –7

Lesson 2-3 Review

1. 11
2. 3
3. 5
4. –10
5. 18
6. –13

Lesson 2-4 Review

1. –18
2. 12
3. –32
4. 120

Lesson 2-5 Review

1. –12
2. 18
3. –3
4. 4

Lesson 2-6 Review

1. 288
2. 564
3. 296
4. 1,357

3

Operations With Fractions

Real numbers belong to one of two camps. A real number can either be written as a ratio of two integers, or it can't. A number that can be written as a ratio of two integers is called a **rational number**. A number that cannot be written as a ratio of two integers is called an **irrational number**. In this chapter we will focus on rational numbers. Irrational numbers will be discussed briefly in Chapter 4.

A rational number is often called a **fraction**, though the definition of a fraction isn't as precise as that of a rational number. Some people use the word *fraction* to describe the ratio of two numbers that may, or may not be, integers. I will use the word *fraction* to mean a rational number. A fraction is written $\frac{a}{b}$, and it means the same thing as $a \div b$. In the fraction $\frac{a}{b}$, the number a is called the **numerator** of the fraction, and the number b is called the **denominator** of the fraction. Remember that when you look at a fraction as a division problem, the number a is called the dividend and the number b is called the divisor. In other words, numerators and dividends represent the same thing, as do denominators and divisors.

3

OPERATIONS WITH FRACTIONS

It is also important to remember that integers, as in the number 3 or the number 7, can be thought of as fractions. In the same way that the sign is *implied* with a positive number, a denominator of 1 is implied with integers.

One of the most interesting fractions is the number 1. It has many forms: It can be written as $\frac{1}{1}, \frac{2}{2}, \frac{3}{3}, \frac{-1}{-1}\ldots$. This is an important property of 1 that we will use repeatedly. In general, any non-zero number divided by itself is 1, so we can represent 1 as the ratio of any non-zero number to itself. In other words, if a represents any non-zero number, then $\frac{a}{a} = 1$.

To find the reciprocal of a fraction, the numerator becomes the denominator and the denominator becomes the numerator. In other words, the **reciprocal** of the fraction $\frac{a}{b}$ is the fraction $\frac{b}{a}$. The reciprocal of an integer is the fraction whose numerator is 1 and whose denominator is equal to the integer. In other words, the reciprocal of a is $\frac{1}{a}$.

Fractions can involve positive or negative numbers in the numerator or the denominator. The following fractions are all equal in value:

$$-\frac{1}{2} = \frac{-1}{2} = \frac{1}{-2}$$

If your fraction is negative, it doesn't matter whether the negative sign is in front of the fraction, associated with the numerator or associated with the denominator. Remember that fractions involve division, and the sign of a quotient will be negative if the divisor and the dividend have opposite signs. Remember that the parts of a fraction correspond to the parts of a quotient.

While I'm on the subject of negative signs, recall that the product of two negative numbers is a positive number, and a negative number divided by a negative number is a positive number. The same rule holds for fractions: If both the numerator and the denominator are negative, then the overall fraction will be positive.

Lesson 3-1: Multiplying Fractions

Multiplying fractions is very straightforward. When multiplying two fractions, multiply the two numerators together to get the numerator of the product and multiply the two denominators together to get the denominator of the product. We can write this algebraically as:

$$\frac{a}{b} \times \frac{c}{d} = \frac{a \times c}{b \times d}$$

Example 1

Find the product: $\dfrac{2}{3} \times \dfrac{5}{11}$

Solution: Multiply the two numerators together and the two denominators together: $\dfrac{2}{3} \times \dfrac{5}{11} = \dfrac{2 \times 5}{3 \times 11} = \dfrac{10}{33}$

Example 2

Find the product: $\dfrac{3}{16} \times \dfrac{5}{7}$

Solution: $\dfrac{3}{16} \times \dfrac{5}{7} = \dfrac{3 \times 5}{16 \times 7} = \dfrac{15}{112}$

Example 3

Find the product: $5 \times \dfrac{2}{21}$

Solution: Turn 5 into a fraction and then follow the rule for multiplying fractions:

$$5 \times \frac{2}{21} = \frac{5}{1} \times \frac{2}{21} = \frac{5 \times 2}{1 \times 21} = \frac{10}{21}$$

Example 4

Find the product: $\dfrac{2}{5} \times \left(\dfrac{-3}{7} \right)$

Solution: The overall product will be negative, but as stated earlier, the negative sign can go in front of the whole fraction:

$$\frac{2}{5} \times \left(\frac{-3}{7} \right) = \frac{2 \times (-3)}{5 \times 7} = \frac{-6}{35} = -\frac{6}{35}$$

Example 5

Find the product: $\dfrac{-3}{4} \times \dfrac{5}{-7}$

Solution: The signs of both the numerator and the denominator will be negative, so the overall fraction will be positive:

$$\frac{-3}{4} \times \frac{5}{-7} = \frac{(-3) \times 5}{4 \times (-7)} = \frac{-15}{-28} = \frac{15}{28}$$

Now that I have worked out some examples, it is time for you to put pencil to paper and try your hand at multiplying fractions.

Lesson 3-1 Review

Find the following products.

1. $\dfrac{3}{5} \times \dfrac{1}{4}$

2. $\dfrac{9}{20} \times \dfrac{3}{7}$

3. $\dfrac{6}{17} \times \dfrac{2}{5}$

4. $\dfrac{8}{11} \times \dfrac{4}{9}$

Lesson 3-2: Simplifying Fractions

Two fractions are equivalent if they represent the same number. For example, the fraction $\frac{12}{3}$ is equivalent to 4. In other words, when you divide 3 into 12 you get 4. Remember: Any whole number can be treated as a fraction with the denominator equal to 1: $\frac{4}{1} = 4 \div 1 = 4$. Sometimes it helps to simplify fractions (or complicate them, depending on the your perspective).

Numbers that are a combination of whole numbers and fractions are called **mixed numbers**. Examples of mixed numbers are $9\frac{2}{3}$ and $4\frac{1}{2}$. Fractions in which the numerator is greater than the denominator are called **improper fractions**. Examples of improper fractions are

$\frac{5}{3}$ and $\frac{10}{2}$. The term *improper* fraction is misleading. There is nothing "improper" about improper fractions; it's just that some people prefer to see improper fractions written as mixed numbers. If you encounter a teacher who abhors improper fractions, my advice is to convert improper fractions to mixed numbers and keep the peace.

The way to convert an improper fraction to a mixed number is to do the division. The result of the division will be a whole number (the quotient) and a remainder; these two numbers (the quotient and the remainder) will appear in the mixed number. The mixed number will be represented as the quotient plus a fraction with the numerator equal to the remainder and the denominator equal to the divisor.

Example 1

Convert the improper fraction $\dfrac{27}{4}$ to a mixed number.

Solution: Do the division:

$$\begin{array}{r} 6 \\ 4\overline{)27} \\ \underline{24} \\ 3 \end{array}$$

So $\dfrac{27}{4} = 6\dfrac{3}{4}$.

Example 2

Convert the improper fraction $\dfrac{310}{7}$ to a mixed number.

Solution: Do the division:

$$\begin{array}{r} 44 \\ 7\overline{)310} \\ \underline{28} \\ 30 \\ \underline{28} \\ 2 \end{array}$$

So $\dfrac{310}{7} = 44\dfrac{2}{7}$.

Mixed numbers may be more aesthetically pleasing to some people, but as far as functionality goes, nothing beats an improper fraction. So it is beneficial to discuss the process of converting a mixed number to an improper fraction. Recall that a mixed number has a whole part and a fractional part. To convert a mixed number to an improper fraction, you will need to find the numerator and denominator of this improper fraction. The numerator of the improper fraction is the result when you take the product of the whole number and the denominator of the mixed number and add it to the numerator of the fractional part of the mixed number. The denominator of the improper fraction will be the same as the denominator of the fraction in the mixed number.

Example 3

Convert $4\dfrac{2}{3}$ to an improper fraction.

Solution: The numerator of the improper fraction will be $4 \times 3 + 2 = 14$, and the denominator of the improper fraction will be 3; the improper fraction is $\dfrac{14}{3}$.

You can always check your answer by doing the division again. If you are able to convert your improper fraction into the original mixed number, you should have confidence in your answer.

Example 4

Convert $5\dfrac{6}{11}$ to an improper fraction.

Solution: The numerator of the improper fraction will be $5 \times 11 + 6 = 61$, and the denominator of the improper fraction will be 11. The improper fraction is $\dfrac{61}{11}$.

Converting between one form and another doesn't really count as a "simplification." One way to truly **simplify fractions** is to pull out factors that are common to both the numerator and the denominator. In order to do this, you must completely factor both the numerator and the denominator. Then use the properties $\frac{a}{a}=1$ and $\frac{a\times c}{b\times d}=\frac{a}{b}\times\frac{c}{d}$. A fraction is reduced when the numerator and the denominator have no common factors other than 1. Remember from Lesson 1-7 that when two numbers have no common factors, they are called relatively prime.

Example 5

Reduce the fraction: $\dfrac{6}{9}$

Solution: The first step in reducing a fraction is to completely factor the numerator and the denominator. $6 = 2 \times 3$ and $9 = 3 \times 3$. Both 6 and 9 are evenly divisible by 3, so we can write:

$$\frac{6}{9}=\frac{2\times 3}{3\times 3}=\frac{3}{3}\times\frac{2}{3}=1\times\frac{2}{3}=\frac{2}{3}$$

As you become familiar with the process, you will find yourself leaving out some of the intermediate steps and writing something similar to this:

$$\frac{6}{9}=\frac{\cancel{3}\times 2}{\cancel{3}\times 3}=\frac{2}{3}$$

This process is often referred to as "canceling." In the example we just did, we "cancelled the 3s." What we are really doing is making use of those useful properties of multiplication mentioned earlier. Remember that the key to reducing fractions is to factor the numerator and the denominator completely. This is done by creating factor trees for the numerator and the denominator. Canceling is done *only* when multiplication and division are the *only* operations involved in the fraction.

Example 6

Reduce the fraction: $\dfrac{12}{30}$

Solution: Factor the numerator and denominator completely: $12 = 2 \times 2 \times 3$ and $30 = 2 \times 3 \times 5$. Both 12 and 30 are evenly divisible by both 2 and 3, so:

$$\frac{12}{30} = \frac{\cancel{2} \times 2 \times \cancel{3}}{\cancel{2} \times \cancel{3} \times 5} = \frac{2}{5}$$

When you multiply two fractions together, the numbers can get awfully large. After you multiply the two fractions together, you will have to check to make sure that the numerator and denominator are relatively prime, which means that you may find yourself factoring large numbers. However, there is a better way. Instead of blindly multiplying numerators together and denominators together, factor both numerators and both denominators, and try to cancel out common factors. You will find that the job will be much easier.

Example 7

Perform the following multiplication: $\dfrac{5}{12} \times \dfrac{63}{130}$

Solution: Rather than finding the products 5×63 and 12×130, it's better to factor the numerators and denominators completely and cancel what you can:

$$\frac{5}{18} \times \frac{63}{130} = \frac{\cancel{5} \times 7 \times \cancel{3} \times \cancel{3}}{2 \times \cancel{3} \times \cancel{3} \times 13 \times 2 \times \cancel{5}} = \frac{7}{52}$$

If you had decided to ignore my earlier advice and multiplied the fractions together without trying to cancel first, you would have gotten the fraction $\frac{315}{2,340}$. Create a factor tree for those numbers and then get back to me!

Example 8

Perform the following multiplication: $3 \times \dfrac{2}{3}$

Solution: $3 \times \dfrac{2}{3} = \dfrac{3}{1} \times \dfrac{2}{3} = \dfrac{\cancel{3} \times 2}{\cancel{3}} = 2$

This last example occurs so frequently that it is worth writing out more formally.

> If a and b are non-zero numbers, then $a \times \left(\dfrac{b}{a} \right) = b$.

The a's cancel out, and you are left with just the numerator of the second factor.

Lesson 3-2 Review

Evaluate the following.

1. Convert $\dfrac{45}{13}$ to a mixed number.

2. Convert $7\dfrac{3}{8}$ to an improper fraction.

3. Reduce $\dfrac{72}{90}$.

4. Find the product $\dfrac{6}{35} \times \dfrac{25}{27}$.

5. Find the product $\dfrac{11}{15} \times \dfrac{25}{33}$.

6. Find the product $\dfrac{5}{36} \times \dfrac{60}{85}$.

Lesson 3-3: Dividing Fractions

As you go through life, you will find that there are plenty of opportunities to divide one integer by another integer. For example, suppose you and three friends order a pizza. You agreed to divide the pie evenly. Suppose that the pizza is cut into eight slices. In order to determine how many pieces you each are allowed to eat, you would need to divide 8 by 4; each of you would be able to eat two slices.

There are also times when you will need to divide one fraction by another fraction. Suppose you have a recipe for cookies that calls for $\frac{2}{3}$ of a cup of butter. If you have 3 cups of butter, you may want to calculate how many batches of cookies you can make without having to go to the grocery store. To figure this out, you would take 3 cups and divide it by $\frac{2}{3}$. Now that you see **why** you would want to divide one fraction by another fraction, we can talk about **how** to divide one fraction by another fraction.

A **complex fraction** is a fraction where the numerator, denominator, or both are themselves fractions. A complex fraction can be converted to a simple fraction by multiplying the numerator by the reciprocal of the denominator:

$$\frac{\frac{a}{b}}{\frac{c}{d}} = \frac{a}{b} \times \frac{d}{c}$$

In other words, when you divide one fraction by another fraction, you invert the denominator and multiply. This is usually shortened to just "invert and multiply," but you have to remember that it is the denominator that gets inverted!

Example 1

Simplify the complex fraction: $\frac{\frac{3}{4}}{\frac{2}{5}}$

Solution:

Invert the denominator and multiply: $\frac{\frac{3}{4}}{\frac{2}{5}} = \frac{3}{4} \times \frac{5}{2} = \frac{15}{8}$

You could also write your answer as a mixed number: $\frac{15}{8} = 1\frac{7}{8}$

Example 2

Simplify the complex fraction: $\frac{\frac{8}{11}}{\frac{4}{33}}$

Solution: Invert the denominator and multiply:

$$\frac{\frac{8}{11}}{\frac{4}{33}} = \frac{8}{11} \times \frac{33}{4} = \frac{\cancel{2} \times \cancel{2} \times 2 \times 3 \times \cancel{11}}{\cancel{2} \times \cancel{2} \times \cancel{11}} = \frac{6}{1} = 6$$

Lesson 3-3 Review

Simplify the following complex fractions. Put your answer in reduced form.

1. $\dfrac{\frac{3}{8}}{\frac{4}{9}}$

2. $\dfrac{\frac{4}{9}}{\frac{14}{27}}$

3. $\dfrac{\frac{7}{15}}{\frac{9}{10}}$

Lesson 3-4: Adding Fractions

If the two fractions that you are adding have the same denominator, then adding the fractions is fairly straightforward: All you do is add the numerators and keep the denominator the same. Of course, once you have added the fractions, you may need to reduce the result. For example, we can find the sum $\frac{4}{15} + \frac{2}{15}$:

$$\frac{4}{15} + \frac{2}{15} = \frac{4+2}{15} = \frac{6}{15} = \frac{2 \times \cancel{3}}{\cancel{3} \times 5} = \frac{2}{5}$$

It's when the denominators are different that things get a bit tricky. The only time you are allowed to add two fractions is when their *denominators are the same*. If the two fractions have different denominators you have to turn them into fractions that have the same denominator. The only tool you have at your disposal is to multiply by 1, but remember that 1 has many disguises.

Suppose you want to add the fractions $\frac{1}{3}$ and $\frac{1}{2}$. Clearly these two fractions have different denominators, so you can't just start adding numerators together. You must first change these fractions so that they have the same denominator, and the best denominator to use is the least common multiple. The least common multiple of 3 and 2 is 6. We will then convert each fraction into a form that has 6 as the denominator without changing the value of the fraction. This can be thought of as the reverse of canceling. Instead of reducing a fraction, you are introducing common factors to *both* the numerator and the denominator so that the denominators in the two fractions being added will be the same.

For 3 to become a 6 in the denominator, multiply 3 by 2, so multiply by 1 disguised as $\frac{2}{2}$.

$$\frac{1}{3} = 1 \times \frac{1}{3} = \frac{2}{2} \times \frac{1}{3} = \frac{2}{6}$$

For 2 to become a 6 in the denominator, multiply 2 by 3, so multiply by 1 disguised as $\frac{3}{3}$.

$$\frac{1}{2} = 1 \times \frac{1}{2} = \frac{3}{3} \times \frac{1}{2} = \frac{3}{6}$$

When you change the denominator of a fraction, it is important to keep the *value* of the fraction the same. The golden rule of changing the denominator of a fraction is to do to the numerator whatever you do to the denominator. When you change the numerator and the denominator by the same *factor*, you are effectively multiplying by 1.

Now that we have written $\frac{1}{3}$ and $\frac{1}{2}$ as fractions with the same denominator, we can add them together by simply adding the numerators:

$$\frac{1}{3} + \frac{1}{2} = \frac{2}{6} + \frac{3}{6} = \frac{5}{6}$$

Once we've added the two fractions we must check whether or not we can reduce the resulting fraction. In this case, 5 and 6 are relatively prime, so there is no further reduction.

Example 1

Find: $\dfrac{3}{25} + \dfrac{9}{20}$

Solution: In order to add these two fractions, we need them to have a common denominator. First, factor both denominators to find the greatest common factor: $25 = 5 \times 5$ and $20 = 2 \times 2 \times 5$ so the greatest common factor is 5. Next, find the least common multiple:

$$\frac{25 \times 20}{5} = \frac{500}{5} = 100$$

So 100 is the common denominator. Change each fraction so that its denominator is 100:

$$\frac{3}{25} = 1 \times \frac{3}{25} = \frac{4}{4} \times \frac{3}{25} = \frac{12}{100}$$

$$\frac{9}{20} = 1 \times \frac{9}{20} = \frac{5}{5} \times \frac{9}{20} = \frac{45}{100}$$

Finally, add the two fractions together:

$$\frac{3}{25} + \frac{9}{20} = \frac{12}{100} + \frac{45}{100} = \frac{57}{100}$$

Check to see if any reduction is possible. Because 57 and 100 are relatively prime, no further reduction is possible, and the problem is solved.

Lesson 3-4 Review

Find the following sums.

1. $\dfrac{27}{62} + \dfrac{19}{62}$

2. $\dfrac{7}{32} + \dfrac{15}{20}$

3. $\dfrac{21}{55} + \dfrac{11}{20}$

Lesson 3-5: Subtracting Fractions

Subtracting fractions is very similar to adding fractions. The denominators of the two fractions must be the same in order to be subtracted. Remember that, just as integers can, fractions can be positive or negative. When you practiced adding and subtracting integers, you had to pay attention to the signs of the integers involved. Subtracting a negative integer was equivalent to adding the corresponding positive integer, because two negatives make a positive. You must pay attention to the signs of the two fractions involved in the subtraction. Subtracting a negative fraction is equivalent to adding the positive fraction.

Example 1

Evaluate: $\dfrac{12}{35} - \dfrac{3}{5}$

Solution: The common denominator is 35. The fraction $\dfrac{12}{35}$ already has a denominator equal to 35, so we only need to change the denominator of the second fraction in the expression:

$$\dfrac{3}{5} \times \dfrac{7}{7} = \dfrac{21}{35}$$

Now we are ready to subtract:

$$\dfrac{12}{35} - \dfrac{3}{5} = \dfrac{12}{35} - \dfrac{21}{35} = -\dfrac{9}{35}$$

Example 2

Evaluate: $\dfrac{15}{22} - \left(-\dfrac{3}{10}\right)$

Solution: Find the common denominator:

$22 = 2 \times 11$ and $10 = 2 \times 5$, so the common denominator is $2 \times 5 \times 11$, or 110. Both of the fractions need to be changed:

$$\dfrac{15}{22} \times \dfrac{5}{5} = \dfrac{75}{110}$$

$$\dfrac{3}{10} \times \dfrac{11}{11} = \dfrac{33}{110}$$

Now we are ready to subtract and reduce:

$$\dfrac{15}{22} - \left(-\dfrac{3}{10}\right) = \dfrac{75}{110} - \left(-\dfrac{33}{110}\right) = \dfrac{75}{110} + \dfrac{33}{110} = \dfrac{108}{110} = \dfrac{\cancel{2} \times 2 \times 3 \times 3 \times 3}{\cancel{2} \times 5 \times 11} = \dfrac{54}{55}$$

Example 3

Evaluate: $\dfrac{5}{12} - \left(\dfrac{2}{-5}\right)$

Solution: Find the common denominator: 12 and 5 are relatively prime, so their common denominator is 60. Both of the denominators need to be changed:

$$\dfrac{5}{12} \times \dfrac{5}{5} = \dfrac{25}{60}$$

$$\frac{2}{5} \times \frac{12}{12} = \frac{24}{60}$$

Now we are ready to subtract:

$$\frac{5}{12} - \left(\frac{2}{-5}\right) = \frac{25}{60} - \left(-\frac{24}{60}\right) = \frac{25}{60} + \frac{24}{60} = \frac{49}{60}$$

Example 4

Evaluate: $-\dfrac{7}{30} - \dfrac{3}{5}$

Solution: Find the common denominator: $30 = 2 \times 3 \times 5$ and 5 is prime, so their common denominator is 30. Only the second fraction's denominator needs to be changed:

$$\frac{3}{5} \times \frac{6}{6} = \frac{18}{30}$$

Now we are ready to subtract:

$$-\frac{7}{30} - \frac{3}{5} = -\frac{7}{30} - \frac{18}{30} = -\frac{25}{30} = -\frac{5 \times \cancel{5}}{2 \times 3 \times \cancel{5}} = -\frac{5}{6}$$

Lesson 3-5 Review

Subtract the following fractions.

1. $\dfrac{3}{5} - \dfrac{1}{3}$

2. $\dfrac{4}{5} - \dfrac{2}{7}$

3. $-\dfrac{3}{10} - \dfrac{4}{7}$

4. $\dfrac{3}{16} - \dfrac{6}{11}$

Lesson 3-6: Fractions and the Distributive Property

The distributive property still applies when you work with fractions. For example, the expressions $\frac{3}{4} \times \left(\frac{3}{8} + \frac{2}{3}\right)$ and $\frac{3}{4} \times \frac{3}{8} + \frac{3}{4} \times \frac{2}{3}$ give the same numerical result. The advantage to using the distributive property is that you may be able to avoid the process of finding a common denominator and adding the fractions together, because multiplication of fractions does not require a common denominator.

Example 1

Use the distributive property to evaluate: $12 \times \left(\dfrac{2}{3} + \dfrac{1}{4} \right)$

Solution: Use the distributive property and simplify:

Use the distributive property instead of trying to add $\frac{2}{3}$ and $\frac{1}{4}$.

$$12 \times \left(\frac{2}{3} + \frac{1}{4} \right) = \frac{12}{1} \times \frac{2}{3} + \frac{12}{1} \times \frac{1}{4}$$

Cancel the common factors in each fraction. Treat each fraction *separately*. You are not allowed to cancel across an addition symbol.

$$\frac{2 \times 2 \times 2 \times \cancel{3}}{\cancel{3}} + \frac{\cancel{2} \times \cancel{2} \times 3}{\cancel{2} \times \cancel{2}}$$

Add the two integers together.

$$8 + 3$$
$$11$$

Lesson 3-6: Review

Use the distributive property to evaluate the following expressions.

1. $10 \times \left(\dfrac{2}{5} + \dfrac{1}{2} \right)$

2. $24 \times \left(\dfrac{1}{6} + \dfrac{2}{3} \right)$

3. $15 \times \left(\dfrac{3}{5} - \dfrac{1}{3} \right)$

4. $20 \times \left(\dfrac{1}{2} - \dfrac{1}{4} \right)$

Answer Key

Lesson 3-1 Review

1. $\frac{3}{20}$

2. $\frac{27}{140}$

3. $\frac{12}{85}$

4. $\frac{32}{99}$

Lesson 3-2 Review

1. $3\frac{6}{13}$

2. $\frac{59}{8}$

3. $\frac{4}{5}$

4. $\frac{10}{63}$

5. $\frac{5}{9}$

6. $\frac{5}{51}$

Lesson 3-3 Review

1. $\dfrac{\frac{3}{8}}{\frac{4}{9}} = \frac{3}{8} \cdot \frac{9}{4} = \frac{27}{32}$

2. $\dfrac{\frac{4}{9}}{\frac{14}{27}} = \frac{4}{9} \cdot \frac{27}{14} = \frac{6}{7}$

3. $\dfrac{\frac{7}{15}}{\frac{9}{10}} = \frac{7}{15} \cdot \frac{10}{9} = \frac{14}{27}$

Lesson 3-4 Review

1. $\frac{23}{31}$

2. $\frac{7}{32} + \frac{15}{20} = \frac{5}{5} \cdot \frac{7}{32} + \frac{15}{20} \cdot \frac{8}{8} = \frac{31}{32}$

3. $\frac{21}{55} + \frac{11}{20} = \frac{4}{4} \cdot \frac{21}{55} + \frac{11}{20} \cdot \frac{11}{11} = \frac{41}{44}$

Lesson 3-5 Review

1. $\frac{4}{15}$

2. $\frac{18}{35}$

3. $-\frac{61}{70}$

4. $-\frac{63}{176}$

Lesson 3-6 Review

1. 9

2. 20

3. 4

4. 5

4

Exponents

Recall that multiplication is shorthand notation for addition: $4 \times 5 = 4 + 4 + 4 + 4 + 4$.

In a similar way, exponents are a shorthand notation for multiplication.

Lesson 4-1: Positive Integer Powers

Exponents represent the number of times that a number is multiplied by itself. For example, the product $4 \times 4 \times 4 \times 4 \times 4$ involves multiplying 4 by itself 5 times. Instead of writing out all of the 4s, we write 4^5. In this expression, the number 5 is called the **exponent**, or the **power**, and the number 4 is called the **base**.

When dealing with exponential expressions, it is important to correctly identify the base and the exponent. If the base is positive, then this task is straightforward. For example, in the expression 3^8, the base is 3 and the exponent is 8. This expression is a shorthand way of expressing the product of 3 times itself 8 times. We can expand 3^8 as $3^8 = 3 \times 3 \times 3 \times 3 \times 3 \times 3 \times 3 \times 3$. Exponents may not seem to be a very important invention, but with them, we can write numbers that are even larger

4 EXPONENTS

than we can comprehend. In general, an exponential expression is written a^n, in which a is the base and n is the exponent. The expression a^n is read "a to the nth power" or "a to the n." By convention, any number a raised to the first power is just that number: $a^1 = a$. Examples of this are $4^1 = 4$ and $8^1 = 8$.

Identifying the base becomes more difficult when *negative* numbers are involved. When the base is a negative number, parentheses become very important. For example, if the base is -4 and the exponent is 6, you would write this number as $(-4)^6$. The parentheses make it absolutely clear that the base is negative. If you leave off the parentheses and write -4^6 what you are actually writing is the number $-(4)^6$. In other words, -4^6 is the negative of 4^6, and 4^6 is the number with base 4 and exponent 6. There is a difference between $(-4)^6$ and -4^6. It's worth expanding both expressions just to make the point:

$$(-4)^6 = (-4)(-4)(-4)(-4)(-4)(-4) = 4{,}096$$
$$-4^6 = -(4)^6 = -(4 \times 4 \times 4 \times 4 \times 4 \times 4) = -4{,}096$$

Notice that the magnitudes of these two numbers are the same, but they have opposite signs. We can determine whether an exponential expression will be positive or negative by trying to observe a pattern. Let's expand powers of -1:

$$(-1)^1 = -1$$
$$(-1)^2 = (-1)(-1) = 1$$
$$(-1)^3 = \left[(-1)(-1)\right](-1) = 1 \times (-1) = -1$$
$$(-1)^4 = \left[(-1)(-1)\right]\left[(-1)(-1)\right] = 1 \times 1 = 1$$
$$(-1)^5 = \left[(-1)(-1)\right]\left[(-1)(-1)\right](-1) = 1 \times 1 \times (-1) = -1$$

When the exponent is even, the answer is positive, and when the exponent is odd the answer is negative. This pattern always holds:

➲ When you multiply an even number of negative numbers together, the result will be a positive number.

➲ When you multiply an odd number of negative numbers together, the result will be a negative number.

The easiest way to determine whether a particular product of numbers is positive or negative is to count the number of negative numbers involved in the product, or look at the exponents of the negative numbers involved in the product. Because $(-4)^6$ involves a negative number being raised to an even power, we would expect the result to be a positive number, which it is.

Example 1

Will the following products be positive or negative?

a. $(-6)^4$

b. -6^4

c. $-2 \times (-4)^6$

d. -2×5^3

e. $5 \times (-3)^6$

f. $-5 \times (-2)^7$

Solution:

a. The overall product will be positive. There are 4 negative signs (because the negative is part of the base). $(-6)^4$ means $(-6)(-6)(-6)(-6)$.

b. The overall product will be negative. There is only one negative sign (because the negative is not part of the base). -6^4 means the *opposite* of 6^4.

c. The overall product will be negative. There are 7 negative signs in the product, one from the -2 and 6 from $(-4)^6$.

d. The overall product will be negative. There is only one negative sign in the product.

e. The overall product will be positive. There are 6 negative signs involved in the product from $(-3)^6$.

f. The overall product will be positive. There are 8 negative signs involved in the product, one from the -5 and 7 from $(-2)^7$.

The introduction of exponents requires an expansion to the order of operations. Exponential expressions involve repetitive multiplication, and multiplication and division are done right after any instructions in parentheses. As a result, exponentiation will come before multiplication or division in the order of operations. Parentheses still come first, though. Our expanded order of operations is now as follows:

1. Parentheses.

2. Exponentiation.

3. Multiplication and division read left to right.

4. Addition and subtraction read left to right.

Example 2

Use the order of operations to evaluate the following expressions:

a. 2×6^2

b. -3×2^4

c. $2 \times (-5)^2$

d. $-6 \times (-2)^3$

Solution:

a. $2 \times 6^2 = 2 \times 36 = 72$

b. $-3 \times 2^4 = -3 \times 16 = -48$

EXPONENTS

4

c.   First, find $(-5)^2$: $(-5)^2 = (-5) \times (-5) = 25$

Then use it to evaluate $2 \times (-5)^2$: $2 \times (-5)^2 = 2 \times 25 = 50$

d.   First, find $(-2)^3$: $(-2)^3 = (-2)(-2)(-2) = -8$

Then use it to evaluate $-6 \times (-2)^3$: $-6 \times (-2)^3 = (-6) \times (-8) = 48$

Again, an exponential expression is written as a^n and spoken as "a to the nth power." It is sometimes read as just "a to the n" or as "the nth power of a." There are some special powers that have specific names. For example, a^2 is read "a squared," and a^3 is read "a cubed." The reason for these names stems from their geometrical interpretation, as we will see in Chapter 10.

Exponents were created to simplify repetitive multiplication. Because of how they got their start, exponential expressions only make sense if the exponent is a positive integer. We will have to carefully expand our interpretation of exponents if we want to work with exponential expressions that involve powers that are either negative integers or zero.

Lesson 4-1 Review

Use the order of operations to evaluate the following expressions.

1. 3×2^4

2. -4×3^2

3. $-5 \times (-2)^3$

Lesson 4-2: Rules for Exponents

Many mathematical rules have been developed by working out specific problems and looking for patterns. Now that we can deal with positive integer exponents, it's time to make some observations. Let's look at what happens when we multiply two exponential expressions with the same base.

For example, let's find the product of 2^3 and 2^5:

$$2^3 \times 2^5 = (2 \times 2 \times 2) \times (2 \times 2 \times 2 \times 2 \times 2)$$
$$= 2 \times 2 \times 2 \times 2 \times 2 \times 2 \times 2 \times 2 = 2^8$$

Notice that the exponents of the two products are 3 and 5, and the exponent of the result is 8, which is 3+5. This observation is actually our **product rule** for exponents: When you multiply two numbers with the same base, you add the exponents. This can be stated mathematically as:

$$a^m \times a^n = a^{m+n}$$

Example 1

Find the following products. Leave your answers as exponential expressions:

a. $6^5 \times 6^7$

b. $(-3)^6 \times (-3)^8$

c. $7^4 \times 7^3$

Solution:

a. $6^5 \times 6^7 = 6^{5+7} = 6^{12}$

b. $(-3)^6 \times (-3)^8 = (-3)^{6+8} = (-3)^{14}$

c. $7^4 \times 7^3 = 7^{4+3} = 7^7$

Next let's explore what happens when you divide two exponential expressions with the same base. For example, let's evaluate $\frac{5^7}{5^4}$:

$$\frac{5^7}{5^4} = \frac{5 \times 5 \times 5 \times \cancel{5} \times \cancel{5} \times \cancel{5} \times \cancel{5}}{\cancel{5} \times \cancel{5} \times \cancel{5} \times \cancel{5}} = 5 \times 5 \times 5 = 5^3$$

Notice that every factor of 5 in the denominator cancels with a 5 in the numerator. The resulting exponent is just what you'd get if you subtracted the exponent in the denominator from the exponent in the numerator. That gives us our **quotient rule** for exponents: When you

divide two numbers with the same base, you subtract the exponent in the denominator from the exponent in the numerator. This can be stated mathematically as:

$$\frac{a^m}{a^n} = a^{m-n}$$

Example 2

Find the following quotients. Leave your answers as exponential expressions:

a. $\dfrac{3^7}{3^5}$

b. $\dfrac{(-4)^9}{(-4)^6}$

c. $\dfrac{5^4}{5^3}$

Solution:

a. $\dfrac{3^7}{3^5} = 3^{7-5} = 3^2$

b. $\dfrac{(-4)^9}{(-4)^6} = (-4)^{9-6} = (-4)^3$

c. $\dfrac{5^4}{5^3} = 5^{4-3} = 5^1 = 5$

We've multiplied and divided exponential expressions. Now it's time to explore what happens when the base of an exponential expression is itself an exponential expression. For example let's examine $\left(6^3\right)^4$. Using our order of operations, we have to work out what is in parentheses first, then deal with the exponent:

$$\left(6^3\right)^4 = \left(6 \times 6 \times 6\right)^4 = \left(6 \times 6 \times 6\right)\left(6 \times 6 \times 6\right)\left(6 \times 6 \times 6\right)\left(6 \times 6 \times 6\right) = 6^{12}$$

Notice that the exponent of the final result is 12, which happens to be the product of the two powers involved in the exponential

4

EXPONENTS

expression: 3 and 4. This generalizes into a **power rule** for exponents: When you raise an exponential expression to a power, you multiply the exponents, or when you raise a power to a power, you multiply the powers. We can write this mathematically as:

$$\left(a^m\right)^n = a^{m \times n}$$

Example 3

Evaluate the following. Leave your answers as exponential expressions:

a. $\left(3^4\right)^5$

b. $\left(\left(-4\right)^9\right)^4$

c. $\left(5^4\right)^3$

Solution:

a. $\left(3^4\right)^5 = 3^{4 \times 5} = 3^{20}$

b. $\left(\left(-4\right)^9\right)^4 = \left(-4\right)^{9 \times 4} = \left(-4\right)^{36}$

c. $\left(5^4\right)^3 = 5^{4 \times 3} = 5^{12}$

Lesson 4-2 Review

Evaluate the following. Leave your answers as exponential expressions.

1. $\left(2^6\right)\left(2^4\right)$

2. $\left(-3\right)^4\left(-3\right)^3$

3. $\dfrac{3^6}{3^2}$

4. $\dfrac{4^8}{4^5}$

5. $\left(\left(-5\right)^4\right)^7$

6. $\left(2^3\right)^6$

Lesson 4-3: Negative Integer Powers

We need an interpretation for when the exponent of a number is negative. Mathematicians have chosen to think of a^{-1} as the reciprocal of a non-zero number a. Recall that we wrote the reciprocal of a non-zero number a as $\frac{1}{a}$. Now we have two ways to express the reciprocal of a: either $\frac{1}{a}$ or a^{-1}.

Using this information, we can interpret exponents that are negative integers. We can use the power rule for exponents to interpret an expression such as 2^{-4}:

$$2^{-4} = \left(2^{-1}\right)^4 = \left(\frac{1}{2}\right)^4 = \left(\frac{1}{2}\right)\left(\frac{1}{2}\right)\left(\frac{1}{2}\right)\left(\frac{1}{2}\right) = \frac{1}{16}$$

On the other hand, we can also interpret 2^{-4} as $\left(2^4\right)^{-1}$. Because $2^4 = 16$, and the reciprocal of 16 is $\frac{1}{16}$, we see that:

$$2^{-4} = \left(2^4\right)^{-1} = 16^{-1} = \frac{1}{16}$$

This works because the power rule involves multiplying exponents, and multiplication is commutative (meaning that the order doesn't matter).

In general, if n is a positive integer, then

$$a^{-n} = \left(a^n\right)^{-1} = \left(a^{-1}\right)^n = \left(\frac{1}{a}\right)^n$$

The role of the negative sign in an exponent is to let you know whether the base belongs in the numerator or the denominator of the fraction. Another way to look at it is that the negative exponent just means to take the reciprocal of the base. Keep in mind that the reciprocal of a is $\frac{1}{a}$, and the reciprocal of $\frac{1}{a}$ is a.

Example 1

Evaluate the following exponential expressions:

a. 3^{-4}

b. $(-4)^{-2}$

c. $\left(\dfrac{1}{5}\right)^{-3}$

d. $\left(-\dfrac{1}{3}\right)^{-3}$

Solution:

a. $3^{-4} = \left(3^4\right)^{-1} = 81^{-1} = \dfrac{1}{81}$

b. $(-4)^{-2} = \left[(-4)^2\right]^{-1} = 16^{-1} = \dfrac{1}{16}$

c. $\left(\dfrac{1}{5}\right)^{-3} = \left[\left(\dfrac{1}{5}\right)^{-1}\right]^3 = 5^3 = 125$

d. $\left(-\dfrac{1}{3}\right)^{-3} = \left[\left(-\dfrac{1}{3}\right)^{-1}\right]^3 = (-3)^3 = -27$

Lesson 4-3 Review

Evaluate the following exponential expressions.

1. 2^{-5}

2. $(-3)^{-3}$

3. $\left(\dfrac{1}{2}\right)^{-3}$

4. $\left(-\dfrac{1}{3}\right)^{-4}$

Lesson 4-4: When the Power Is Zero

In order to interpret an exponential expression that has 0 as the exponent, we must revisit division of exponential expressions. Consider the ratio $\dfrac{a^n}{a^n}$, where $a \neq 0$. Using one of the properties of multiplication discussed previously, any number divided by itself is equal to 1, so:

$$\dfrac{a^n}{a^n} = 1$$

Using the rules for dividing exponential expressions, when you divide two numbers with the same base, you subtract the exponents, so:

$$\frac{a^n}{a^n} = a^{n-n} = a^0$$

In mathematics, consistency is crucial. There's not much of a choice for how to interpret an exponential expression that has 0 as the exponent: $a^0 = 1$. Here is yet another special property of 0: Any number raised to the power of 0 (other than 0, of course) is 1. For example, $4^0 = 1$ and $8^0 = 1$.

Using this idea, we can now explore the exponential expression $\frac{1}{a^{-n}}$:

$$\frac{1}{a^{-n}} = \frac{a^0}{a^{-n}} = a^{0-(-n)} = a^n$$

The main idea here is that when you move an exponential expression from the numerator to the denominator, or vice versa, the net effect is that the sign of the exponent is changed. This is demonstrated by:

$$\frac{1}{4^{-3}} = 4^3 \text{ and } 8^{-4} = \frac{1}{8^4}.$$

Lesson 4-5: Powers of Quotients and Products

Exponential expressions can get even more complicated. Let's examine the expression $(5a)^4$. Whenever you need to simplify exponential expressions, it is always best to start with the meaning of exponents, and then make use of the properties of multiplication. For example,

$$(5a)^4 = (5a)(5a)(5a)(5a) = 5\times5\times5\times5\times a\times a\times a\times a = 5^4 \times a^4$$

Notice that when you raise a product (in this case, the product of 5 and a) to a power, in effect, what you need to do is raise each term involved in the product to that power. It doesn't matter how many terms are involved in the product, or even if the terms in the products are themselves exponential expressions. Each term gets raised to that power. In general, this rule is written as:

$$(a\times b)^n = a^n \times b^n$$

4

EXPONENTS

Example 1

Expand the following products:

a. $(4a)^2$

b. $(2b^2)^4$

c. $(-3a)^5$

Solution:

a. $(4a)^2 = 4^2 \times a^2 = 16a^2$

b. $(2b^2)^4 = 2^4 \times (b^2)^4 = 16b^{2\times4} = 16b^8$

c. $(-3a)^5 = (-1)^5 (3)^5 (a^5) = -243a^5$

Dealing with raising quotients to a power isn't any different. For example:

$$\left(\frac{3}{a}\right)^4 = \left(\frac{3}{a}\right)\left(\frac{3}{a}\right)\left(\frac{3}{a}\right)\left(\frac{3}{a}\right) = \frac{3^4}{a^4}$$

When you raise a quotient to a power, raise the numerator and the denominator to that power. It doesn't matter how complicated the numerator and denominator are. Use the appropriate rules to keep simplifying until there's nothing more you can do. If you take it one step at a time, you should be fine. In general, this rule is written:

$$\left(\frac{a}{b}\right)^n = \frac{a^n}{b^n}$$

Example 2

Evaluate the following:

a. $\left(\dfrac{a}{3}\right)^2$

b. $\left(\dfrac{2b^3}{3}\right)^5$

c. $\left(-\dfrac{4}{5a}\right)^3$

4 EXPONENTS

Solution:

a. $\left(\dfrac{a}{3}\right)^2 = \dfrac{a^2}{3^2} = \dfrac{a^2}{9}$

b. $\left(\dfrac{2b^3}{3}\right)^5 = \dfrac{\left(2b^3\right)^5}{3^5} = \dfrac{2^5 \times \left(b^3\right)^5}{243} = \dfrac{32b^{3\times5}}{243} = \dfrac{32b^{15}}{243}$

c. $\left(-\dfrac{4}{5a}\right)^3 = (-1)^3 \dfrac{4^3}{(5a)^3} = -\dfrac{64}{5^3 \times a^3} = -\dfrac{64}{125a^3}$

Lesson 4-5 Review

Evaluate the following.

1. $(2b)^3$

2. $\left(6a^2\right)^{-3}$

3. $\left(-\dfrac{4}{5a}\right)^3$

4. $\left(\dfrac{-3a}{b}\right)^2$

5. $\left(\dfrac{-3a}{b}\right)^2$

6. $\left(\dfrac{2}{3b}\right)^{-3}$

Lesson 4-6: Square Roots and Irrational Numbers

Squaring a number involves multiplying a number by itself. For example, the square of 3 is $3^2 = 9$; the square of -3 is $(-3)^2 = 9$. Notice that squaring a positive number gives a positive number, and squaring a negative number also results in a positive number. The square of 0 is $0^2 = 0$. If you square any real number, whether it is positive, negative, or 0, you will get a number that is either greater than 0 or equal to 0, but never less than 0. In other words, if a is any number, then $a^2 \geq 0$. There is no way that the square of a real number can be negative.

4

EXPONENTS

Finding the square root of a number involves reversing the process: if $b^2 = a$, then b is called the **square root** of a. For example, because $3^2 = 9$, then 3 is the square root of 9. Notice that $(-3)^2 = 9$ so -3 is also the square root of 9. In fact, all positive numbers have two square roots: a *positive* square root, called the **principal square root**, and a *negative* square root. The two square roots have the same magnitude, but have opposite signs.

Square roots are written using a "square root" symbol: $\sqrt{}$. This symbol is called a radical. The number underneath the **radical** is called the **radicand**. To write things more compactly, sometimes the square roots of a number are written together as $\pm\sqrt{}$. We could write the two square roots of 9 as ± 3. Be careful with this notation and the terminology. If we write $\sqrt{9}$ we are limiting ourselves to the positive square root of 9, but keep in mind that 9 has two square roots: $\pm\sqrt{9}$. The radical by itself refers to the principal (or positive) square root.

Notice that $\sqrt{3^2} = \sqrt{9} = 3$. We can generalize this observation as follows: If a is a positive number, then $\sqrt{a^2} = a$. Remember that this rule only holds if a is a positive number.

Simplifying a square root involves factoring the radicand and pulling out the "perfect squares." **Perfect squares** are numbers whose square roots are integers. For example, because $3^2 = 9$ we write $\sqrt{9} = 3$. Because the square root of 9 is 3 (a whole number), we say that 9 is a perfect square. It's easy to list the perfect squares by squaring the natural numbers. The first 10 perfect squares are 1, 4, 9, 16, 25, 36, 49, 64, 81, and 100. To simplify a radical, you need to factor the radicand, look for perfect squares and make use of the rule that if a is positive, then $\sqrt{a^2} = a$.

Square roots have some very useful properties that are helpful in simplifying square roots. If a and b are positive numbers, then $\sqrt{a \cdot b} = \sqrt{a} \cdot \sqrt{b}$. We can use this relationship to simplify square roots by pulling out the perfect squares: $\sqrt{a^2 \cdot b} = \sqrt{a^2} \cdot \sqrt{b} = a\sqrt{b}$

Example 1

Simplify: $\sqrt{81}$

Solution: Because 81 is a perfect square (you can see it in the list: $9^2 = 81$), we can write $\sqrt{81} = 9$.

Example 2

Simplify: $\sqrt{50}$

Solution: Examine whether or not 50 has any perfect squares as factors. Because $50 = 25 \cdot 2$ we can write:
$$\sqrt{50} = \sqrt{25 \cdot 2} = \sqrt{25} \times \sqrt{2} = 5\sqrt{2}$$

Example 3

Simplify: $\sqrt{120}$

Solution: Factor 120 into products of perfect squares:
$120 = 4 \cdot 30$ so $\sqrt{120} = \sqrt{4 \cdot 30} = \sqrt{4} \times \sqrt{30} = 2\sqrt{30}$

The easiest way to approach these problems is to look at the list of perfect squares and check to see if any of them divide the radicand evenly. If so, factor the radicand into a product of a perfect square and another factor:

radicand = perfect square × other factor

Now do the same thing with the other factor: See if any other perfect squares divide that other factor. If they do, keep factoring. If not, pull out all of the perfect squares that you found.

The only way for the square root of a number to be a whole number is if it is a perfect square, and the only way that the square root of a number can be a rational number is if it is the ratio of two perfect squares. The square root of any other kind of number is an *irrational number*. Recall that irrational numbers are numbers that cannot be written as the ratio of two integers.

4

EXPONENTS

Lesson 4-6 Review

Simplify the following radicals.

1. $\sqrt{64}$

2. $\sqrt{45}$

3. $\sqrt{48}$

4. $\sqrt{216}$

Answer Key

Lesson 4-1 Review

1. 48

2. −36

3. 40

Lesson 4-2 Review

1. 2^{10}

2. $(-3)^7$

3. 3^4

4. 4^3

5. $(-5)^{28}$

6. 2^{18}

Lesson 4-3 Review

1. $\frac{1}{32}$

2. $-\frac{1}{27}$

3. 8

4. 81

Lesson 4-5 Review

1. $8b^3$

2. $\frac{1}{216a^6}$

3. $16a^{12}$

4. $\frac{9a^2}{b^2}$

5. $\frac{256}{b^4}$

6. $\frac{27b^3}{8}$

Lesson 4-6 Review

1. 8

2. $3\sqrt{5}$

3. $4\sqrt{3}$

4. $6\sqrt{6}$

5

Decimals

We use a very clever system to write whole numbers: the base ten system. There are other number systems, such as the Roman numeral system, which is great for keeping track of the number of Super Bowls that have been played. However, there is a reason that the Roman numeral system is not widely used today. Try multiplying XXIX and XIV together using only Roman numerals. The Roman numeral system works well for counting, but not well for adding or multiplying.

In the base ten system we have ten symbols at our disposal: 0, 1, 2, 3, 4, 5, 6, 7, 8, and 9. The number 2,365 represents 2 thousands, 3 hundreds, 6 tens, and 5 ones. The order in which the numbers appear *matters*. As you move to the left, the power of ten increases by 1: the ones' place can be thought of as 10^0, the tens' place is 10^1, the hundreds' place is 10^2, and so on. We can represent any whole number using this system.

Base ten is great, but you should realize that 10 is not the only base that can be used to represent numbers. There are other popular systems that are in use today, such as base two, base eight, and base sixteen, but that is a topic for another day.

Right now, it is just important to understand that everything we have been talking about so far involves the base ten number system.

Lesson 5-1: The Decimal System

The base ten system works well for representing integers, but it doesn't work as well for other types of numbers, such as irrational numbers and many fractions. Fractions and irrational numbers have an integral part, which is an integer, and a fractional part, or a part whose magnitude is less than 1. We use the base ten system to represent the integral part, and we use the decimal system to represent the fractional part. A decimal number is a number with both an integral part (remember that 0 is an integer) and a fractional part. The decimal system is similar to the base ten system because we use the same ten symbols, powers of 10 play an important role in interpreting the numbers, and the order in which you write the numbers in matters. You have been working with decimals for at least as long as you have been working with money. A quarter can be represented as 25 cents or $0.25, but, either way, you know that a quarter is less than a dollar.

The decimal representation of a number involves a decimal point; we use the symbol "." (a period) to represent the decimal point. Numbers to the *left* of the decimal point make up the integral part of the number, and numbers to the *right* make up the fractional part, or decimal fraction. *A decimal number* has two parts: the integral part and the decimal fraction. The first place to the right of the decimal is called the tenths' place. Whatever number is located in the tenths' place is interpreted as being the numerator of a fraction that has 10 as the denominator. The next place to the right is the hundredths' place. Whatever number is located in the hundredths place is interpreted as being the numerator of a fraction that has 100 as its denominator. In our example of a quarter, or $0.25, the 2 is in the tenths' place and the 5 is in the hundredths' place. The 2 in the tenths' place is really $\frac{2}{10}$, and the 5 in the hundredths' place is really $\frac{5}{100}$. Continue interpreting

the places to the right of the decimal point: There is the thousandths'
place, the ten-thousandths' place, and so on. All of these individual
fractions will add up to equal the decimal fraction.

To illustrate the components of a decimal number, consider the
number 12.325. The decimal number 12.325 represents 1 ten, 2 ones,
3 tenths, 2 hundredths, and 5 thousandths. The sum of these components is 12.325:

$$10 + 2 + \frac{3}{10} + \frac{2}{100} + \frac{5}{1000} = 12.325$$

The decimal number 0.3 is equivalent to the fraction $\frac{3}{10}$. If the
magnitude of a decimal number is less than 1, then a 0 is put to the
left of the decimal point to make it clear that the integral part of the
number is 0.

Every integer can be represented as a decimal number by putting
a decimal point immediately after the number in the ones' place. The
number 7 can be written as 7.0 or 7.00 or 7.000. Just as a positive sign
is implied most of the time and is not written, with *integers,* the decimal point is implied most of the time and is not written. Any decimal
representation can be made longer by stringing 0s after the *last digit
to the right* of the decimal point. Although 7, 7.00, and 7.000 have the
same mathematical *value,* they have a different scientific *interpretation.* The number of decimal places indicates the level of accuracy of
a measurement.

One of the most challenging aspects of writing this chapter is that
there is a problem with our terminology. I have been using the words
fraction and *rational number* interchangeably, because using the word
fraction instead of *rational number* has filtered into common usage.
Technically, a rational number is a ratio of two *integers* (such as $\frac{5}{10}$),
and a fraction is a ratio of two *numbers* that do not have to be integers
(such as $\frac{\sqrt{2}}{3}$). We call the part of a decimal number that is less than 1
the *decimal fraction* because it can be *interpreted* as a ratio of two numbers, but it is *written* using the decimal system. You must be careful to

read which kind of fraction you are working with: a *decimal fraction* (such as 0.5), or a *fraction* (such as $\frac{5}{10}$, which is what we have been using to mean a rational number).

Converting from a *decimal fraction* to a rational number (or a traditional fraction) involves converting each digit in the decimal representation into its corresponding fraction and then adding all of the fractions together. Remember that fractions should be written in reduced form.

Example 1

Convert 0.5 to a fraction in reduced form.

Solution: $0.5 = \dfrac{5}{10}$, and $\dfrac{5}{10} = \dfrac{\cancel{5}}{2 \times \cancel{5}} = \dfrac{1}{2}$

Example 2

Convert 0.35 to a fraction in reduced form.

Solution: Write each digit as a fraction and add the fractions together. After you have added the fractions, be sure to check for common factors to cancel:

$$0.35 = \frac{3}{10} + \frac{5}{100} = \frac{30}{100} + \frac{5}{100} = \frac{35}{100} = \frac{\cancel{5} \times 7}{2 \times 2 \times 5 \times \cancel{5}} = \frac{7}{20}$$

Notice that in adding the two fractions together it was necessary to get a common denominator. The common denominator used in this situation was 100, so you had to multiply the fraction $\frac{3}{10}$ by $\frac{10}{10}$. As an intermediate step, the fraction obtained was $\frac{35}{100}$: $0.35 = \frac{35}{100}$

One of the reasons that base ten is so nice to work with is that it is easy to write a decimal number with a fixed number of digits as a fraction that is not necessarily in reduced form. The number of digits to the right of the decimal point will be the power of 10 that appears in the denominator of the fraction. Then, the digits in the decimal fraction go into the numerator. The decimal fraction 0.57 has two digits to the right of the decimal point, so the denominator is 10^2, and 57 goes in the numerator. The decimal fraction 0.57 is written as the fraction $\frac{57}{100}$. The decimal fraction 0.387 has three digits to the right of the decimal point, so the denominator is 10^3, and 387 goes in the numerator. The decimal fraction 0.387 is written as the fraction $\frac{387}{1,000}$. The number of places past the decimal point is the same as the number of 0s in the denominator. That makes converting from a decimal fraction to a fraction very easy. The resulting fraction may need to be reduced.

The reverse process, converting from a fraction to a decimal number, depends on the denominator of the fraction. If the denominator of the fraction being converted is a power of 10, the conversion to a decimal number is as easy as reversing the process we just talked about. The power of 10 in the denominator tells you how many digits in the numerator need to be to the right of the decimal point. The fraction $\frac{523}{100}$ has 10^2 in the denominator, so two of the digits in 523 need to be to the right of the decimal point, and $\frac{523}{100} = 5.23$. This should make sense because you can see that the numerator is larger than the denominator, and $\frac{523}{100}$ is an improper fraction.

If the denominator of the fraction being converted is not a power of 10, then converting from a fraction to a decimal number involves expanding on our method of division. In Chapter 1 we discussed long division and the concept of a remainder. The remainder was the number "left over" when the divisor did not divide evenly into the dividend. With decimal numbers, division doesn't stop with a remainder, it keeps going, and going, and going, until a pattern is discovered. Remember that all *rational numbers* have a decimal representation that either terminates or establishes a pattern. Converting a fraction to a decimal number involves long division until you reach the point of finding the remainder. At this point, instead of stopping, you insert a decimal point and a 0 to the right of the decimal point and continue with the long division process. The decimal point serves to divide the integral part and the fractional part.

DECIMALS

5

Example 3

Convert $\dfrac{117}{20}$ to a decimal number.

Solution:

$$
\begin{array}{r}
5.85 \\
20\overline{)117} \\
10 \\
\hline
17.0 \\
16.0 \\
\hline
1.00 \\
1.00 \\
\hline
0 \\
\end{array}
$$

If we were writing $\frac{117}{20}$ as a *mixed number*, we would have stopped when the result of the subtraction was 17, and our answer would have been $5\frac{17}{20}$. Because we were converting $\frac{117}{20}$ to a *decimal number,* we had to keep going until the division process terminated with a remainder of 0, or until a pattern was established. In this case, the division process terminated with a remainder of 0.

Lesson 5-1 Review

Convert the following decimal fractions to reduced fractions, and the improper fractions to decimal numbers.

1. 0.68

2. 0.12

3. $\dfrac{411}{25}$

4. $\dfrac{47}{40}$

Lesson 5-2: Adding and Subtracting Decimal Numbers

Decimal numbers can be added together in several ways. One way to add decimal numbers is to convert them first to fractions and then add the resulting fractions. The resulting fraction can then be converted to a decimal number using division, if necessary. This method will always work, but after a few examples you will notice some shortcuts.

Example 1

Find the sum: 0.65 + 0.13

Solution: Convert each decimal fraction to a fraction, and then add the fractions together. There is no need to reduce the fractions, because you will have to "un-reduce" them to get a common denominator so that you can add the fractions together:

$$0.65 + 0.13 = \frac{65}{100} + \frac{13}{100} = \frac{78}{100}$$

There is no need to reduce the resulting fraction. You can easily convert it to a decimal fraction because the denominator is a power of 10:

$$\frac{78}{100} = 0.78$$

Adding decimal fractions together is quite straightforward. We just saw how decimal fractions can be converted to fractions that have a denominator equal to a power of 10. Because both decimal numbers will be converted to fractions whose denominators are a power of 10 (though not necessarily the *same* power of 10), the common denominator will also be a power of 10. To determine the power of 10 of the common denominator, simply pick the larger of the two.

Example 2

Find the sum: 0.325 + 0.4062

Solution: Convert each decimal number to a fraction. Notice that 0.325 has three digits to the right of the decimal point (meaning that the denominator of its corresponding fraction would be 10^3) and 0.4062 has four digits to the right of the decimal point (meaning that the denominator of its corresponding fraction will be 10^4). Because $10^4 > 10^3$, 10^4 will be our common denominator:

$$0.325 + 0.4062 = \frac{325}{1,000} + \frac{4,062}{10,000}$$

We can now make the two denominators the same and add the fractions:

$$\frac{325}{1,000} + \frac{4,062}{10,000} = \frac{10}{10} \times \frac{325}{1,000} + \frac{4,062}{10,000} = \frac{3,250}{10,000} + \frac{4,062}{10,000} = \frac{7,312}{10,000}$$

Finally, we will convert our result to a decimal number:

$$\frac{7,312}{10,000} = 0.7312$$

Another way to add two decimal fractions is to line up the decimal points and add the numbers directly. Remember that you can always add 0s after the last digit to the right of the decimal point as placeholders. That means that the equation

$$0.325 + 0.4062 = 0.7312$$

is equivalent to the equation

$$0.3250 + 0.4062 = 0.7312$$

and lining up the digits gives us:

$$
\begin{array}{r}
0.3250 \\
+\ 0.4062 \\
\hline
0.7312
\end{array}
$$

This observation will generalize into a method for adding two decimal numbers together. When you add decimal numbers together, line up the decimal points and put in any necessary 0s needed as placeholders after the last digit to the right of the decimal point. This ensures that both numbers have the same number of digits to the right of the decimal point. Now you are ready to add, right to left, column by column, carrying over when necessary, as we did with whole numbers in Chapter 1.

Example 3

Find the sum: 12.35 + 1.9

Solution: Line up the decimals and write this sum in column form.

$$\begin{array}{r} 12.35 \\ +\ 1.9 \\ \hline \end{array}$$

Because the first decimal number has two places to the right of the decimal point, we need to put a 0 as a placeholder after the 9 in the second decimal number. Now add right to left and carry over when necessary:

$$\begin{array}{r} \overset{1}{1}2.35 \\ +\ 1.90 \\ \hline 14.25 \end{array}$$

Subtracting decimals proceeds similarly. You need to line up the decimal points and fill in any necessary 0s as placeholders so that both decimal numbers have the same number of digits to the right of the decimal point. Then follow the rules for subtracting, and borrow when necessary.

Example 4

Evaluate the expression: 12.35 – 8.22

Solution:
$$\begin{array}{r} 12.35 \\ -\ 8.22 \\ \hline 4.13 \end{array}$$

Example 5

Evaluate the expression: 12.8 – 2.38

Solution: Remember to put the 0 in as a placeholder:

$$\begin{array}{r} 12.\,{}^{7}\cancel{8}\,{}^{1}0 \\ -\ 2.\ \ 3\ \ 8 \\ \hline 10.\ \ 4\ \ 2 \end{array}$$

Example 6

Evaluate the expression: 12.85 − 9.3

Solution:

Remember to put the 0 in as a placeholder:

$$\begin{array}{r} 12.85 \\ -\ 9.30 \\ \hline 3.55 \end{array}$$

Lesson 5-2 Review

Evaluate the following.

1. 39.85 + 24.97

2. 1.55 + 3.2

3. 25.3 + 18.86

4. 25.86 − 14.32

5. 24.46 − 16.8

6. 44.8 − 16.34

Lesson 5-3: Multiplying Decimal Numbers

Multiplying two decimal numbers is very similar to multiplying whole numbers. The first step is to ignore the decimal points, and multiply the numbers together as if they were both whole numbers. To put the decimal point in the right place, count how many digits there are to the *right* of the decimal points in *both* factors. That is how many digits will be to the right of the decimal point in the resulting product.

Example 1

Find the product: 3.2 × 4.6

Solution: First, ignore the decimal points and find the product:

$$\begin{array}{r} 32 \\ \times\ 46 \\ \hline 192 \\ 1{,}280 \\ \hline 1{,}472 \end{array}$$

The whole number product is 1,472. Now we need to place the decimal point. Each factor has one digit to the right of the decimal point, so there are two digits total to the right of the decimal point. We need our product to have two digits to the right of the decimal point, giving an answer of 14.72.

Example 2

Find the product: 3.24×2.8

Solution: First, ignore the decimal points and find the product:

$$
\begin{array}{r}
324 \\
\times\ \ 28 \\
\hline
2{,}592 \\
6{,}480 \\
\hline
9{,}072
\end{array}
$$

The whole number product is 9,072. Now we need to place the decimal point. The first factor has two digits to the right of the decimal point and the second factor has one digit to the right of the decimal point; there are a total of three digits to the right of the decimal point. The resulting product will have three digits to the right of the decimal point, giving an answer of 9.072.

Problems involving multiplying by a power of 10 are the easiest multiplication problems to solve. The effect of multiplying a decimal number by a power of 10 is that only the decimal point moves.

First, let's consider positive powers of 10: numbers such as 10, 100, or 1,000. In this case, the problem becomes one of moving the decimal point *to the right* by one, two, or three places, depending on the power of 10 being multiplied. For example:

$3.2 \times 10 = 3.\underset{\smile}{2} \times 10^{1} = 32$

$3.2 \times 0.01 = 00\underset{\smile\smile}{3.2} \times 10^{-2} = 0.032$

$3.2 \times 1000 = 3.\underset{\smile\smile\smile}{200} \times 10^{3} = 3,200$

If you run out of 0s, you can always put some in, as long as it is to the *right* of the decimal point.

Negative powers of 10, such as 10^{-1}, 10^{-2}, and 10^{-3}, are easy to represent as decimal numbers. For example:

$$10^{-1} = \frac{1}{10} = 0.1$$

$$10^{-2} = \frac{1}{100} = 0.01$$

$$10^{-3} = \frac{1}{1000} = 0.001$$

Multiplying a decimal number by a negative power of 10, such as 0.1, 0.01, or 0.001, involves moving the decimal point to the *left* by the *magnitude* of the power of 10. For example:

$$3.2 \times 0.1 = 3.2 \times 10^{-1} = 0.32$$

$$3.2 \times 0.01 = 003.2 \times 10^{-2} = 0.032$$

$$3.2 \times 0.001 = 0003.2 \times 10^{-3} = 0.0032$$

Lesson 5-3 Review

Find the following products.

1. 5.2×4.1

2. 8.6×3.7

3. 4.58×9.2

4. 4.58×1000

5. 3.24×0.01

6. 8.95×0.1

Lesson 5-4: Dividing Decimal Numbers

Dividing decimals is very similar to dividing whole numbers. Our first goal is to turn the decimal number in the denominator into a whole number. To do this, first determine the number of digits there are to the right of the decimal point in the denominator. Move the decimal points in both the numerator and the denominator over to the right

by that many places, filling in 0s as a placeholder in the numerator if necessary. For example, if we wanted to evaluate $\frac{4.35}{0.87}$ we would first observe that the denominator has two digits to the right of the decimal point. We would then move the decimal points of both the numerator and the denominator to the right by two places, and our problem becomes $\frac{435}{87}$. This is equivalent to multiplying both the numerator and the denominator by 100 (we are multiplying the fraction by the number 1 disguised as the fraction $\frac{100}{100}$). Once we have a whole number in the denominator, we can do the division:

$$
\begin{array}{r}
5 \\
87\overline{)435} \\
\underline{435} \\
0
\end{array}
$$

So, $\dfrac{4.35}{0.87} = 5$.

Example 1

Evaluate: $\dfrac{6.21}{1.38}$

Solution: The decimal number in the denominator has two digits to the right of the decimal point, so we need to move the decimal points in both the numerator and the denominator two places to the right:

$$\frac{6.21}{1.38} = \frac{621}{138}$$

Now do the division, converting this fraction into a decimal number:

$$
\begin{array}{r}
4.5 \\
138\overline{)621.} \\
\underline{552.} \\
69.0 \\
\underline{69.0} \\
.0
\end{array}
$$

So, $\dfrac{6.21}{1.38} = 4.5$.

DECIMALS

5

Example 2

Evaluate: $\dfrac{7.5}{1.25}$

Solution: The decimal number in the denominator has two digits to the right of the decimal point, so we need to move the decimal points in both the numerator and the denominator two places to the right. The numerator only has one digit to the right of the decimal point, so we will need to introduce a 0 as a placeholder in the numerator:

$$\frac{7.5}{1.25} = \frac{750}{125}$$

Notice that the numerator and denominator are whole numbers with at least one common factor (both 750 and 125 are evenly divisible by 5). At this point, we could either factor both the numerator and the denominator and cancel common factors, or we could do the division. I prefer to factor and cancel:

$$\frac{750}{125} = \frac{2 \times 3 \times \cancel{5} \times \cancel{5} \times \cancel{5}}{\cancel{5} \times \cancel{5} \times \cancel{5}} = 6$$

So, $\dfrac{7.5}{1.25} = 6$.

Lesson 5-4 Review

Evaluate the following.

1. $\dfrac{7.4}{1.85}$

2. $\dfrac{3.25}{1.3}$

3. $\dfrac{8.48}{2.65}$

Lesson 5-5: Rounding and Estimating

Sometimes exact numbers give too much information. For example, there may be 7,853 people living in a particular town, but the town's Website may round up to the nearest thousand and say that there are

8,000 residents. People usually prefer nice, round numbers, and the more 0s involved in the number, the happier they are. It doesn't matter that our national debt is currently $8,017,113,351,645. It's enough to say that our national debt is around $8 trillion. The key to rounding is to recognize that 5 is the cut-off. Any digit less than 5 (meaning 0, 1, 2, 3, and 4) turn into 0s. This is called **rounding down**. Any digit greater than or equal to 5 (meaning 5, 6, 7, 8, and 9) turn into 0s *and* bump the immediate digit to the left up by one unit. This is the same thing as turning into a 10 and being carried over. This is called **rounding up**.

The number 852.654 becomes 1,000 when rounding to the nearest thousand. It becomes 900 when rounding to the nearest hundred. It becomes 850 when rounding to the nearest ten. It becomes 853 when rounding to the nearest unit. It becomes 852.7 when rounding to the nearest tenth. It becomes 852.65 when rounding to the nearest hundredth. Rounding depends on the *place* on which you are focusing. When rounding to the nearest ten, your rounding digit is the digit in the tens place. You look at the digit immediately to the right and follow the rounding rules with a cut-off of 5.

When you round a number, you are changing that number. The rounded number is not equal to the original number, but it is close. A handy symbol to represent being close, or approximate, but not equal to, is $\approx$.

Example 1

Round 65.32 to the nearest unit.

Solution: $65.32 \approx 65$

Example 2

Round 4.8562 to the nearest tenth.

Solution: $4.8562 \approx 4.9$

DECIMALS 5

Example 3

Round 4.265 to the nearest unit.

Solution: $4.265 \approx 4$

Rounding enables us to estimate answers without getting bogged down in the details. It can be used as a quick check to see if our answers make sense. For example, suppose that I purchase a sweater for $24.99, a pair of jeans for $42.99, and a shirt for $25.50. To calculate the exact amount of my purchases, I would have to do the addition, down to the penny. This calculation is not particularly difficult, but if I don't have a pencil and a piece of paper handy, I may end up holding up the checkout line. I don't want to just trust the sales clerk, because he or she could have keyed in the wrong price. I'm pretty careful with my hard-earned money. If my goal is to check whether or not the amount I am being charged is correct, I can accomplish that goal by rounding these prices to the nearest dollar and quickly adding these numbers together. My sum should be *close* to the amount that the sales clerk asks for. If the sales clerk wants $140 for these purchases, I can quickly verify whether or not I am being charged the right amount. Depending on how accurate I wanted to be, I could just round to the nearest dollar and find the sum 25 + 43 + 26. I could make it easier and round to the nearest ten dollars and find the sum 20 + 40 + 30. Rounding to the nearest ten dollars, the total should be around $90, so the charge of $140 is too far off the mark. Rounding to the nearest one dollar, the total should be around $94, so still, the charge of $140 is too far off the mark. By doing a little estimating, I could save myself around $50.

Rounding can make numbers a little easier to work with. The calculation is easier, but the trade-off is accuracy. When you balance your checkbook, you want to be accurate, but when you are budgeting expenses, you can estimate your monthly phone bill, food bill, and so forth. If you need to be accurate, don't round and don't estimate. If you want to check your work and see if you are in the ballpark, round and estimate.

You can round and estimate with multiplication as well as addition. For example, to estimate the product 7.65 × 3.3, you could round 7.65 to 8 and 3.3 to 3: 8 × 3 = 24. So 7.65 × 3.3 is roughly 24. The exact answer is 25.245. We're a little off, but we're still in the neighborhood. Keep in mind that if you round down, your estimate will be a little low, and if you round up, your estimate will be a little high.

Example 1

Scott charges $2.10 per square foot to tile a floor. If he charges me $750 to tile a 300-square-foot room, is he charging me the right amount?

Solution: Round $2.10 to the nearest dollar: $2.1 \approx 2$. Scott should have charged me roughly 300 × 2 or $600 to tile the room; $750 is way too high! He definitely overcharged me for the job.

Example 2

To celebrate their first softball victory of the season, Michael wants to treat his friends to pizza. If each pizza costs $7.50 and Michael wants to buy 8 pizzas, how much money will he spend?

Solution: Round the price of each pizza to the nearest dollar: $7.50 \approx \$8$, so 8 pizzas will cost roughly 8 × 8 or $64.

Example 3

Estimate the product: 0.088 × 40.48

Solution: Round 0.088 to the nearest tenth: $0.088 \approx 0.1$. Round 40.48 to the nearest unit: $40.48 \approx 40$. The product 0.088 × 40.48 will be close to the product 0.1 × 40 = 4.

So, $0.088 \times 40.48 \approx 4$.

Estimating quotients is similar to estimating products. Round the numerator and the denominator and then do the division. The place you round to depends on the numbers you are working with and how accurate you want to be.

Example 4

Estimate the quotient: $\dfrac{32.5}{5.28}$

Solution: Round the numerator to the nearest ten: $32.5 \approx 30$. Round the denominator to the nearest unit: $5.28 \approx 5$.

So, $\dfrac{32.5}{5.28} \approx \dfrac{30}{5} = 6$.

Lesson 5-5 Review

Round to the nearest unit and estimate the following.

1. $23.65 + 19.75 + 22.56$

2. $55.67 - 23.94$

3. 14.89×5.22

4. 5.26×3.05

5. $\dfrac{55.24}{4.85}$

Lesson 5-6: Comparing Decimals and Fractions

Comparing two *integers* involves determining which of the two numbers is larger than the other. One way to compare two integers is to place them carefully on the number line. The number that is farthest to the *right* is the larger of the two numbers. On the other hand, the number that is farthest to the *left* is the smaller of the two numbers. Now, we need a method for comparing fractions or decimal numbers.

To compare two decimal numbers, start by comparing their integral parts. If the two decimal numbers have the same integral part, start comparing their fractional parts. To compare the fractional parts, start comparing the digits immediately to the right of the decimal point and keep going right until one number wins.

Example 1

Compare: 10.56 and 8.37

Solution: Compare the integral parts of the two numbers: The integral part of 10.56 is 10, and the integral part of 8.37 is 8. Clearly 10.56 wins: $10.56 > 8.37$.

Example 2

Compare: 1.369 and 1.375

Solution: Both numbers have the same integral part, so they tie, and we will need to compare the fractional parts. Start with the tenths' place: Both numbers have a 3 in the tenths' place, so they still tie. 1.369 has a 6 in the hundredths' place and 1.375 has a 7 in the hundredths' place, so 1.375 wins: $1.369 < 1.375$.

Example 3

Compare: 1.25 and 1.254

Solution: Both numbers have the same integral part, so they tie, and we will need to compare the fractional parts. Start with the tenths' place: Both numbers have a 2 in the tenths' place, so they still tie. Both numbers have a 5 in the hundredths' place, so they continue to tie. Now compare the digits in the thousandths' place: 1.25 has a 0 in the thousandths' place and 1.254 has a 4 in the thousandths' place, so 1.254 wins: $1.25 < 1.254$.

Comparing two fractions is a little bit more tricky. If the denominators of the two fractions are equal, then the problem involves comparing the numerators.

Example 4

Compare: $\dfrac{3}{31}$ and $\dfrac{8}{31}$

Solution: Because both denominators are the same, we need only compare the numerators.

Because $3 < 8$, $\dfrac{3}{31} < \dfrac{8}{31}$.

When the denominators are different, you can take two approaches. The first approach involves finding the lowest common denominator and then comparing numerators. The second approach involves converting each fraction to a decimal number and comparing the results.

Example 5

Compare: $\dfrac{2}{3}$ and $\dfrac{3}{4}$

Solution: In this case, the lowest common denominator is 12. Transform both fractions so that their denominators are 12 and then compare their numerators:

$$\frac{2}{3} \times \frac{4}{4} = \frac{8}{12} \qquad \frac{3}{4} \times \frac{3}{3} = \frac{9}{12}$$

Because $8 < 9$, $\dfrac{8}{12} < \dfrac{9}{12}$, so $\dfrac{2}{3} < \dfrac{3}{4}$.

Example 6

Compare: $\dfrac{7}{20}$ and $\dfrac{9}{25}$

Solution: We could find the lowest common denominator and then compare numerators. I'd like to illustrate the second approach here, just for a change.

First, convert each fraction to a decimal number:

$$
\begin{array}{r}
.35 \\
20\overline{)7.0} \\
\underline{6.0} \\
1.00 \\
\underline{1.00} \\
0
\end{array}
\qquad
\begin{array}{r}
.36 \\
25\overline{)9.0} \\
\underline{7.5} \\
1.50 \\
\underline{1.50} \\
0
\end{array}
$$

So, $\dfrac{7}{20} = .35$ and $\dfrac{9}{25} = .36$.

Now compare the decimal representations. Both numbers have a 3 in the tenth's place, so they tie.

$\dfrac{7}{20}$ has a 5 in the hundredth's place and

$\dfrac{9}{25}$ has a 6 in the hundredth's place, so $\dfrac{9}{25}$ wins.

Therefore $\dfrac{7}{20} < \dfrac{9}{25}$.

We can also compare a fraction and a decimal number. For this comparison, you can either convert the fraction to a decimal number and then compare the two decimal numbers, or you can convert the decimal number to a fraction, and then compare the two fractions. You will get the same result regardless of the method you choose.

Example 7

Compare $\dfrac{4}{5}$ and 0.805.

Solution: Convert 0.805 to a fraction and compare the two fractions:

$$0.805 = \frac{805}{1000} = \frac{\cancel{5} \times 7 \times 23}{2 \times 2 \times 2 \times \cancel{5} \times 5 \times 5} = \frac{161}{200}$$

In order to compare the two fractions, find the lowest common denominator. In this case, the lowest common denominator is 200, so I only have to change $\dfrac{4}{5}$:

$$\frac{4}{5} = \frac{40}{40} \times \frac{4}{5} = \frac{160}{200}$$

Because $161 > 160$, the decimal number wins:

$$\frac{4}{5} < 0.805$$

Lesson 5-6 Review

Compare the following numbers.

1.   1.542 and 1.562

2.   10.89 and 10.82

3. $\dfrac{13}{20}$ and $\dfrac{19}{25}$

4. $\dfrac{5}{7}$ and $\dfrac{7}{9}$

5   DECIMALS

5

DECIMALS

Answer Key

Lesson 5-1 Review

1. $\frac{17}{25}$

2. $\frac{3}{25}$

3. 16.44

4. 1.175

Lesson 5-2 Review

1. 64.82

2. 4.75

3. 44.16

4. 11.54

5. 7.66

6. 28.46

Lesson 5-3 Review

1. 21.32

2. 31.82

3. 42.136

4. 4,580

5. 0.0324

6. 0.895

Lesson 5-4 Review

1. 4

2. 2.5

3. 3.2

Lesson 5-5 Review

1. 67

2. 32

3. 75

4. 15

5. 11

Lesson 5-6 Review

1. $1.542 < 1.562$

2. $10.89 > 10.82$

3. $\frac{13}{20} < \frac{19}{25}$ (convert to decimal numbers)

4. $\frac{5}{7} < \frac{7}{9}$ (get a common denominator and compare fractions)

Expressions and Equations

Problems in algebra often involve variables. I told you that a variable is a letter than can be used to represent *any* number, and I have used variables each time I have given you a generalized rule. Another way to use variables is to have them represent an unknown quantity.

Variables often appear in algebraic expressions. An **algebraic expression** is a statement that combines numbers and variables together using some of the operations we have discussed: addition, subtraction, multiplication, division, and exponentiation. For example, the expression $x + 10$ means that you take the number x (whatever it happens to be) and add 10 to it. The variable x could represent the amount of money in your pocket, and 10 could represent the amount of money that your neighbor is going to give you for cleaning up her yard. If we specify a value for x, say $x = 15$, meaning that you have $15 in your pocket, then we can evaluate $x + 10$, which would be 25 (or $15 + 10$).

There is no rule that says that a variable must be a specific letter. Any letter will do. We could even use words, such as *cost*, *money*, and *pounds*, but words are a little awkward, which is why we usually just use the first letter of the word as our variable. So if we wanted a variable to represent weight, we might reach for a "w," and if we wanted a variable to represent distance we might reach for a "d." Mathematicians are pretty boring and usually just grab the first variable off the shelf: x. Because the variable x and the multiplication symbol $\times$ look very similar, I will use a different symbol for multiplication to avoid any confusion. I will use the $\cdot$ or $(\ )(\ )$.

Lesson 6-1: Evaluating Expressions

In order to evaluate an expression for particular values of the variables, replace the variables with their values and do the math. This process is often referred to as "plug and chug" because you are plugging in the numbers and chugging away at the calculation. Make sure that you pay attention to the order of operations.

Sometimes multiplication symbols are omitted in expressions. For example, the expression $3x$ represents the action of taking the variable x and multiplying it by 3. The multiplication symbol is implied. Remember that multiplication is shorthand for addition, so $3x$ actually means $x + x + x$. The number 3 in front of the variable is called the **coefficient** of the variable. This will come in handy when we start manipulating expressions.

Example 1

Evaluate the expression $3x + 5$ when $x = 4$.

Solution: Everywhere you see an x, write a 4. Then work through the calculation.

When $x = 4$, $3x + 5 = 3 \cdot 4 + 5 = 12 + 5 = 17$

Example 2

Evaluate the expression $6x + 10$ when $x = -2$.

Solution: Everywhere you see an x, write a -2. Then work through the calculation.

When $x = -2$, $6x + 10 = 6 \cdot (-2) + 10 = -12 + 10 = -2$.

Example 3

Evaluate the expression $(2y) \div (y + 8)$ when $y = -6$.

Solution: Everywhere you see a y, write a -6. Because y is a negative number, I will put parentheses around it to keep it with the 6. Now, work through the calculation.

When $y = -6$,

$(2y) \div (y + 8) = (2 \cdot (-6)) \div (-6 + 8) = (-12) \div 2 = -6$.

When division is involved, writing the expression out using the symbol $\div$ becomes somewhat awkward. You will usually see expressions such as $(2y) \div (y + 8)$ written as $\frac{2y}{y+8}$. In this fractional form, you must be careful with the order of operations. The expression $\frac{2y}{y+8}$ should really be thought of as $\frac{(2y)}{(y+8)}$; and be sure to perform the operations in parentheses before doing the division.

Expressions can involve more than one variable. In order to evaluate expressions that involve more than one variable, you will be given specific values for all of the variables.

Example 4

Evaluate the expression $\dfrac{a+2b}{a-b}$ when $a = 4$ and $b = 1$.

Solution: Everywhere there is an a, plug in 4, and where there's a b, plug in a 1.

When $a = 4$ and $b = 1$, $\dfrac{a+2b}{a-b} = \dfrac{(a+2b)}{(a-b)} = \dfrac{4+2\cdot1}{4-1} = \dfrac{6}{3} = 2$.

Example 5

Evaluate the expression $x^2 + 5$ when $x = 4$.

Solution: When $x = 4$, $x^2 + 5 = 4^2 + 5 = 16 + 5 = 21$.

Example 6

Evaluate the expression $(a + b)(a - b)$ when $a = 5$ and $b = 2$.

Solution: When $a = 5$ and $b = 2$,

$(a+b)(a-b) = (5+2)(5-2) = 7 \cdot 3 = 21$

6 EXPRESSIONS AND EQUATIONS

Lesson 6-1 Review

Evaluate the following expressions when $x = 5$ and $y = -3$.

1. $3x + 2$

2. $4x + 2y$

3. $x^2 + 2y$

4. $(x + 2y)(x - y)$

5. $\dfrac{x+3}{y-2}$

6. $\dfrac{2y}{3x}$

Lesson 6-2: Equations

An **equation** is a statement that says that two expressions are equal. The expressions that make up an equation are separated by the symbol "=," which is called an **equal sign**. Usually equations will contain variables, but they don't have to.

You have already seen several examples of equations in this book. The statements $2^3 = 2 \cdot 2 \cdot 2$ and $a^m \cdot a^n = a^{m+n}$ are two examples of equations that you may recall reading in Chapter 4. The first equation does not contain any variables, and the second equation contains three variables: a, n, and m.

Equations can have numbers and variables on both sides. Equations can be simple or complicated. The important thing about an equation is that whatever expression appears on the left-hand side of the equal sign is *equivalent* to whatever expression appears on the right-hand side. The two expressions may not look alike, but they must represent the same thing.

When you are given an equation with only one variable, your job will be to "solve the equation." To solve an equation, you will need to determine the numerical value or values of the variable that make the equation true. This process involves moving all of the numbers to one side of the equation and all of the terms involving variables to the other side. When moving around the terms in an equation, you must follow some rules.

Lesson 6-3: Equality

The concept of equality is an important idea and has very special properties worth examining in more detail. It is used to relate seemingly distinct objects together and is called a relation. A **relation** is something that compares two objects. For example, if I had 10 one-dollar bills and you had a ten-dollar bill, then we would have an equal amount of money, even though the form of our money is different. If my son broke his piggy bank and discovered that he had 40 quarters, he would have the same amount of money as we do. This example helps illustrate three important properties of equality:

> ⊃ Equality is **reflexive**. This means that any object is equal to itself. This idea is almost self-evident. To illustrate the reflexive property mathematically, we write $a = a$. For example, $5 = 5$.

> ⊃ Equality is **symmetric**. This means that if object A is equal to object B, then object B is equal to object A. This property emphasizes the fact that the order in which we equate objects doesn't matter. It doesn't matter if object A appears on the left- or the right-hand side of the $=$. To state this idea mathematically we write: If $a = b$, then $b = a$. For example, if $3 \cdot 4 = 12$, then $12 = 3 \cdot 4$.

> ⊃ Equality is **transitive**. This means that if object A is equal to object B, and object B is equal to object C, then object A is equal to object C. Object B is just an intermediate object that can be eliminated in the comparison; we can cut to the chase and set object A equal to object C. This can be expressed mathematically as: If $a = b$ and $b = c$, then $a = c$. For example, if $10 \cdot 5 = 50$ and $50 = 25 \cdot 2$, then $10 \cdot 5 = 25 \cdot 2$.

Because these three properties are so important in mathematics, we say that any relation that has all three of these properties is an **equivalence relation**. It is important to note that *every* equivalence relation has all three of these properties. There are other relations besides equality that we will use later on that are not equivalence relations. Those relations will be missing at least one of the three required properties.

Lesson 6-4: Algebraic Properties of Equality

Equations represent balance, and any change that you make to an equation must maintain that balance. Maintaining that balance means that there are only certain things that you are allowed to do to an equation. The key point is that whatever you do to one side of the equation you must also do to the other side of the equation. You are not allowed to tip the scales and favor one side over another side.

The first algebraic property of equality is known as the **addition property of equality**. It states that you are allowed to add the same amount to both sides of an equation. Mathematically, we write this property as follows:

If a, b, and c are numbers with $a = b$,
then $a + c = b + c$.

We are not allowed to favor one side of the equality over the other side; the quantity c was added to both sides, so the balance is maintained. Looking back on the example with money, if you have a ten-dollar bill and I have 10 one-dollar bills, then we each have the same amount of money. If a kind stranger gives us each a twenty-dollar bill, then we are both still equally rich; we both have a total of $30. Notice that in the addition property of equality, the signs of a, b, and c were not specified. That's because it doesn't matter. If I have 10 one-dollar bills and my son has 40 quarters, and we both decide to spend $2 on an ice-cream cone, then we will both end up with $8. The addition property of equality is important because it allows us to add (or subtract) the same number from both sides of an equation without changing the validity of that statement.

The second algebraic property of equality is known as the **multiplication property of equality**. It states that you are allowed to multiply both sides of an equality by the same amount. Mathematically, we write this property as follows:

> If a, b, and c are numbers with $a = b$,
> then $(a)(c) = (b)(c)$.

Again, one side is not favored over the other side; both sides are treated equally because both sides are being multiplied by c. Let's go back to our example with money. If you have a ten-dollar bill and I have 10 one-dollar bills, and we both double our money in a bet, then we both now have $20. Notice again that with this property I never specified the signs of a, b, or c. I also never specified whether c was greater than 1 or less than 1. This property says that I can multiply (or divide) both sides of an equation by the same number (and this number can be positive or negative, or it can be a fraction, a decimal number, or an integer) and not change the validity of the statement.

These properties are certainly interesting in their own right, but, more importantly, they will play a crucial role in your success in being able to solve algebraic equations. We'll see how they are used in the next section.

Lesson 6-5: Solving Linear Equations in 1 Step

An equation involves setting two algebraic expressions equal to each other and solving for the variable in the expressions. A **linear expression** contains only constants and one variable raised to the first power. A linear expression can look something such as $3x + 2$, $3(x - 2) + 1$, or $(2x + 3) + (5x - 2)$. Common to all of these expressions is the fact that there is only one variable, which is raised to the first power. The third expression has two terms that involve the variable, but the only variable in the expression is x.

A **linear equation** is an equation in which two linear expressions are set equal to each other. We can use the algebraic properties of equality to help us solve linear equations.

There are key transformations that you will use repeatedly when solving equations. When you use these transformations, you will produce an equation that has the same solutions as the original equation, but the solutions will be easier to see. In other words, you will be transforming the original equation into a different, equivalent equation that, ideally, will be easier to solve. The goal is to write an equation of the form $x =$ a number.

The first four transformations that we will practice are described in the following table:

Transformation	Original Equation	Transformation	New Equation
Interchange the sides of the equation	$4 = x$	Interchange	$x = 4$
Simplify one or both sides	$2x - x = 10 - 6$	Simplify	$x = 4$
Add the same number to each side and simplify	$x - 5 = 3$	Add 5 to both sides and simplify	$x - 5 + 5 = 3 + 5$ $x = 8$
Subtract the same number from each side and simplify	$x + 7 = 20$	Subtract 7 from both sides and simplify	$x + 7 - 7 = 20 - 7$ $x = 13$

Anything that is done to one side of the equation must also be done on the other side of the equation. This ensures that balance is maintained. Remember that the goal is to determine the value of the variable. In order to solve for the variable, you must get the variable by itself. These transformations help you isolate the variable.

Example 1

Solve the equation: $x - 5 = 12$

Solution: In order to isolate the variable, we need to get rid of the 5. To do this, just add 5 to both sides and simplify:

$$x - 5 = 12$$

Add 5 to both sides. $x - 5 + 5 = 12 + 5$

Simplify. $x = 17$

The last step is to check our work.
Substitute $x = 17$ into our original equation and simplify.
Is $17 - 5$ equal to 12?
Yes. We can be confident that our answer is correct.

Example 2

Solve the equation: $x + 8 = -5$

Solution: Subtract 8 from both sides and simplify:

$$x + 8 = -5$$

Subtract 8 from both sides. $x + 8 - 8 = -5 - 8$

Simplify. $x = -13$

Let's check our work.
Substitute $x = -13$ into the original equation.
Is $-13 + 8$ equal to -5? Yes.

Transforming equations using addition and subtraction is just the beginning. We can also transform equations using multiplication and division. You will need to transform equations using multiplication and division whenever the coefficient in front of the variable is any number other than 1. Remember that the goal is to get the variable by itself so that the equation explicitly states what the variable is equal to.

6 EXPRESSIONS AND EQUATIONS

Transformation	Original Equation	Transformation	New Equation
Multiply each side of the equation by the same number	$\dfrac{1}{2}x = 8$	Multiply both sides of the equation by 2 and simplify	$2 \cdot \left(\dfrac{1}{2}x\right) = 2 \cdot 8$ $x = 16$
Divide each side of the equation by the same number	$3x = 15$	Divide both sides of the equation by 3 and simplify	$\dfrac{3x}{3} = \dfrac{15}{3}$ $x = 5$

You can also look at the division transformation in terms of multiplication. If you start with the equation $3x = 15$, you can multiply both sides of the equation by the reciprocal of the coefficient in front of the variable. In this case, the coefficient in front of the variable is 3, and its reciprocal is $\frac{1}{3}$. So, in order to solve the equation $3x = 15$, we could multiply both sides of this equation by $\frac{1}{3}$:

$$3x = 15$$
$$\frac{1}{3} \cdot (3x) = \frac{1}{3} \cdot 15$$
$$x = 5$$

This perspective is helpful when the coefficient in front of the variable is not an integer.

Example 3

Solve the equation: $20 = -\dfrac{1}{4}x$

Solution: You may have gotten used to seeing the variable on the left rather than on the right. Don't let that bother you; switch them around using the transformation that lets you interchange the sides of the equation if it makes you feel better. Then multiply both sides of the equation by –4 (the reciprocal of $-\frac{1}{4}$) and simplify:

$$20 = -\frac{1}{4}x$$

Interchange the sides of the equation. $-\frac{1}{4}x = 20$

Multiply both sides of the equation by –4. $(-4) \cdot \left(-\frac{1}{4}x \right) = (-4) \cdot 20$

Simplify. $x = -80$

The last step is to check our work.

Is $\left(-\frac{1}{4} \right) \cdot (-80)$ equal to 20? Yes, it is.

Example 4

Solve the equation: $4x = -9$

Solution: Divide both sides of the equation by 4 and simplify:

$$4x = -9$$

Divide both sides of the equation by 4. $\dfrac{4x}{4} = \dfrac{-9}{4}$

Simplify. $x = \dfrac{-9}{4} = -2\dfrac{1}{4}$

Of course you could have just multiplied both sides of the equation by $\frac{1}{4}$. The last step is to check our work.

Is $4 \cdot \left(-\dfrac{9}{4} \right)$ equal to –9? Yes.

Example 5

Solve the equation: $-\dfrac{4}{5}x = 8$

Solution: Multiply both sides of the equation by the reciprocal of $-\frac{4}{5}$ (which is $-\frac{5}{4}$) and simplify:

$$-\frac{4}{5}x = 8$$

Multiply both sides of the equation by $-\frac{5}{4}$. $\left(-\dfrac{5}{4} \right) \cdot \left(-\dfrac{4}{5}x \right) = \left(-\dfrac{5}{4} \right) \cdot 8$

Simplify. $x = -10$

Finally, check our work.

Is $\left(-\dfrac{4}{5}\right) \cdot (-10)$ equal to 8? Yes.

Lesson 6-5 Review

Solve the following equations.

1. $x + 2 = 9$
2. $x - 3 = 9$
3. $8 = -\dfrac{1}{3}x$
4. $3x = -4$
5. $-\dfrac{2}{5}x = 4$

Lesson 6-6: Solving Linear Equations in 2 Steps

In general, when solving a linear equation involving one variable the first thing you need to do is move all of the terms that involve the variable over to one side of the equation, and move all of the terms that don't involve the variable over to the other side of the equation. If, after combining all of the terms together, the coefficient in front of the variable is a number other than 1, you will need to multiply both sides of the equation by the reciprocal of the coefficient in front of the variable. Once you have done that, you should have an explicit equation for what numerical value the variable has to be. The last step is to check your work (using the original problem statement) to make sure that your answer is correct. This last step is the one that is most often skipped, but it's one that you really should get into the habit of doing. By checking your work, you will know whether your answer is correct or not. If it's not, you'll have to start over.

Example 1

Solve the equation: $\dfrac{1}{3}x + 5 = -3$

Solution: To isolate the variable, you must move the 5 that appears on the left side of the equation over to the right side. This can

EXPRESSIONS AND EQUATIONS

6

be done by subtracting 5 from both sides of the equation and simplifying:

$$\frac{1}{3}x + 5 = -3$$

Subtract 5 from both sides.

$$\frac{1}{3}x + 5 - 5 = -3 - 5$$

Simplify.

$$\frac{1}{3}x = -8$$

Next, we want the coefficient in front of the variable to be 1. You can transform this equation by multiplying both sides of this equation by 3 and then simplifying.

$$\frac{1}{3}x = -8$$

Multiply both sides by 3.

$$3 \cdot \left(\frac{1}{3}x\right) = 3 \cdot (-8)$$

Simplify.

$$x = -24$$

Finally, check your answer.

Is $\frac{1}{3} \cdot (-24) + 5 = -8 + 5$ equal to -3? Yes.

Example 2

Solve the equation: $\frac{1}{3}x = \frac{3}{5}x - 2$

Solution: In this case, you will see terms with variables appearing on both sides of the equation. You want to collect all of the terms with variables on one side of the equation and then solve the equation using the same techniques described earlier. In order to get all of the variables on one side of the equation, we need to subtract $\frac{3}{5}x$ from both sides and simplify:

$$\frac{1}{3}x = \frac{3}{5}x - 2$$

Subtract $\frac{3}{5}x$ from both sides.

$$\frac{1}{3}x - \frac{3}{5}x = \frac{3}{5}x - 2 - \frac{3}{5}x$$

Simplify.

$$\frac{1}{3}x - \frac{3}{5}x = -2$$

Multiply both fractions by 1.

$$\left(\frac{5}{5}\right)\frac{1}{3}x - \left(\frac{3}{3}\right)\frac{3}{5}x = -2$$

Make the denominators the same.

$$\frac{5}{15}x - \frac{9}{15}x = -2$$

Simplify.

$$-\frac{4}{15}x = -2$$

Now that our variables are on one side and the numbers are on the other side, we can solve the equation. Multiply both sides of the equation by $-\frac{15}{4}$ and simplify:

$$-\frac{4}{15}x = -2$$

Multiply both sides by $-\frac{15}{4}$.

$$\left(-\frac{15}{4}\right)\left(-\frac{4}{15}\right)x = \left(-\frac{15}{4}\right)(-2)$$

Simplify.

$$x = \frac{30}{4} = \frac{15}{2} = 7\frac{1}{2}$$

Write your answer as either an improper fraction or a mixed number.

The last step is to check our work.

When $x = \frac{15}{2}$, the left-hand side of the equation gives $\frac{1}{3} \times \frac{15}{2} = \frac{5}{2}$

and the right-hand side of the equation gives

$$\frac{3}{5} \times \frac{15}{2} - 2 = \frac{9}{2} - 2 = \frac{9}{2} - \frac{4}{2} = \frac{5}{2}.$$

Both sides calculate to $\frac{5}{2}$, so our solution is correct.

Example 3

Solve the equation: $5x = 2x + 6$

Solution: Gather the variables on one side of the equation and the numbers on the other side. Then solve the equation:

$$5x = 2x + 6$$

Subtract $2x$ from both sides.

$$5x - 2x = 2x + 6 - 2x$$

Simplify.

$$3x = 6$$

Multiply both sides of the equation by $\frac{1}{3}$.

$$\frac{1}{3} \cdot (3x) = \frac{1}{3} \cdot 6$$

Simplify.

$$x = 2$$

Finally, we'll check our answer: If $x = 2$, then the left-hand side of the equation gives $5x = 5 \times 2 = 10$ and the right-hand side gives $2x + 6 = 2 \times 2 + 6 = 4 + 6 = 10$. Both sides calculate to 10, so our answer is correct.

There are potential problems that you have to look out for when you are working with an equation that has more than one term involving the variable. One of the potential problems is that a solution to the equation may not exist. For example, the equation $x = x + 1$ has no solution, because there is no number that is equal to itself plus 1 (not even 0 can do that!). If you subtract x from both sides of the equation and simplify, you get:

$x = x + 1$

$x - x = x + 1 - x$

$0 = 1$

Because this last equation is absurd, so is our original equation. Absurd equations have no solution.

Lesson 6-6 Review

Solve the following equations.

1. $\frac{2}{3}x - 3 = 8$

2. $10x = 3x - 14$

3. $4x - 5 = -1$

4. $\frac{2}{5}x = \frac{1}{2}x - 1$

6

EXPRESSIONS AND EQUATIONS

Answer Key

Lesson 6-1 Review

1. 17

2. 14

3. 19

4. −8

5. $-\frac{8}{5}$

6. $-\frac{2}{5}$

Lesson 6-5 Review

1. Subtract 2 from both sides: $x = 7$

2. Add 3 to both sides: $x = 12$

3. Multiply both sides by −3: $x = -24$

4. Divide both sides by 3: $x = -\frac{4}{3}$

5. Multiply both sides by $-\frac{5}{2}$: $x = -10$

Lesson 6-6 Review

1. Add 3 to both sides and then multiply by $\frac{3}{2}$: $x = \frac{33}{2}$

2. Subtract $3x$ from both sides and then divide both sides by 7: $x = -2$

3. Add 5 to both sides and then divide both sides by 4: $x = 1$

4. Subtract $\frac{1}{2}x$ from both sides and then multiply both sides by 10: $x = 10$

Ratios, Proportions, and Percents

A **ratio** is a fraction that is used to compare two quantities. Ratios involve division. There are several ways to write a ratio. You can write the ratio 5 out of 10 as "5 to 10," "5:10," or $\frac{5}{10}$. If you play the lottery or enter a raffle, the odds that you will win are often given as a ratio. The closer the ratio is to one, the better the chances are that you will win.

Lesson 7-1: Ratios and Unit Rates

A **rate** is a ratio that compares two quantities with different units. Rates are used to provide information in a variety of situations. For example, the rate at which people drive on an interstate is a ratio of the number of miles driven to the time that has passed. This is usually referred to as the speed that you are traveling. A vehicle's gas mileage is the ratio of the number of miles driven to the volume of gasoline consumed. The density of a substance is the ratio of the mass of the substance to the volume that the substance occupies. Higher odds mean that an event has a better chance of happening.

Rate	Fraction
Driving Speed	$\dfrac{\text{miles driven}}{\text{elapsed time}}$
Gas Mileage	$\dfrac{\text{miles driven}}{\text{volume of gasoline consumed}}$
Density	$\dfrac{\text{mass}}{\text{volume}}$

A **unit rate** is a rate that has a denominator equal to 1. Usually speed, gas mileage, and density are stated in terms of a unit rate, such as miles per hour, miles per gallon, and grams per cubic centimeter.

Example 1

While shopping for detergent, you notice that the store carries a 10-ounce box for $2.59, an 18-ounce box for $4.39, and a 32-ounce box for $6.72. Which box of detergent gives you the most detergent for the least amount of money?

Solution: Because the boxes of detergent have different weights and different prices, the best thing to do is compare unit rates of dollars to ounces of detergent. Find the unit rates for all three sizes.

Size	Unit Rate
10 ounces (small)	$\dfrac{\$2.59}{10 \text{ ounces}} = \dfrac{\$0.259}{\text{ounce}}$
18 ounces (medium)	$\dfrac{\$4.39}{18 \text{ ounces}} = \dfrac{\$0.244}{\text{ounce}}$
32 ounces (large)	$\dfrac{\$6.72}{32 \text{ ounces}} = \dfrac{\$0.210}{\text{ounce}}$

Now it is clear that an ounce of detergent costs $0.259 if you buy the small box, $0.244 if you buy the medium box, and $0.210 if you buy the large box. You will spend less money per ounce of detergent if you purchase the large box.

Instead of comparing dollars to ounces, you could have compared ounces to dollars. Then your ratios would have told you how much detergent you would be purchasing for $1. In that case, the best deal would have been the one that resulted in the largest ratio (the most detergent for $1), as shown in the following table. It doesn't matter which way your ratios are calculated, as long as they are all calculated the same way. The important thing is to think about *what* your ratios mean and base your decision accordingly:

Size	Unit Rate
10 ounces (small)	$\dfrac{10 \text{ ounces}}{\$2.59} = \dfrac{3.86 \text{ ounces}}{\text{dollar}}$
18 ounces (medium)	$\dfrac{18 \text{ ounces}}{\$4.39} = \dfrac{4.10 \text{ ounces}}{\text{dollar}}$
32 ounces (large)	$\dfrac{32 \text{ ounces}}{\$6.72} = \dfrac{4.76 \text{ ounces}}{\text{dollar}}$

As a word of warning, you cannot always assume that a larger size has a lower unit rate. There are many times that I have calculated the unit rates of different sizes of products such as detergent, dog food, and charcoal, and found that the larger size had a more expensive corresponding unit rate. It may be that the items were incorrectly priced, or that the discrepancy resulted from a temporary price reduction in the smaller-sized item. I have learned to always compare unit rates when deciding between different sizes.

Lesson 7-1 Review

Answer the following.

1. When pricing dog food, I discovered that a 5-pound bag of food costs $8.49, a 20-pound bag costs $21.49, and a 40-pound bag costs $29.99. Calculate the unit rates for each of these bags.

2. When pricing soda, I discovered that a 12-pack costs $2.99 and a 24-pack costs $5.00. Calculate the unit rates for each of these options.

3. Kendelyn bought a new car. When she drove 200 miles in the city, the car consumed 8 gallons of gasoline. When she drove 300 miles on the interstate, the car consumed 10 gallons of gasoline. Find the gas mileage for the city and the interstate driving conditions.

4. On her cross-country trip to Wisconsin, Betty drove 980 miles in 14 hours. Find her speed as a unit rate.

Lesson 7-2: Proportions

In the last chapter, we solved linear equations. Remember that a linear *equation* is an equation created by setting two linear *expressions* equal to each other. In the last section, we talked about a different kind of algebraic expression: a ratio. When two ratios are set equal to each other, we have an equation that is called a proportion. A **proportion** is an equality between two ratios:

$$\frac{a}{b} = \frac{c}{d}$$

Proportions can be simplified by multiplying both sides of the equality by the product of the denominators and simplifying using the properties of multiplication:

$$\cancel{b}d \times \left(\frac{a}{\cancel{b}}\right) = b\cancel{d} \times \left(\frac{c}{\cancel{d}}\right)$$

$$ad = bc$$

This process is often described as **"cross-multiplying"** and can be visualized as:

$$\frac{a}{b} \diagup\!\!\!\!\!\diagdown \frac{c}{d}$$

The products *ad* and *cd* are called the **cross-products** of the proportion.

If a proportion involves a variable, you can cross-multiply and then solve for the variable as we did in the last chapter.

Example 1

Solve the proportion: $\dfrac{x}{16} = \dfrac{15}{40}$

Solution: There are two ways to solve this proportion. The first way involves noticing that this proportion also happens to be a linear equation. As such, we can solve this equation by using the techniques discussed in Chapter 6. Multiply both sides of the equation by 16 and reduce the resulting fraction:

$$16 \times \frac{x}{16} = 16 \times \frac{15}{40}$$

$$x = \frac{16 \times 15}{40} = \frac{\cancel{2} \times \cancel{2} \times \cancel{2} \times 2 \times 3 \times \cancel{5}}{\cancel{2} \times \cancel{2} \times \cancel{2} \times \cancel{5}} = 6$$

The other way to solve this proportion is to cross-multiply, and then solve for the variable:

$$\frac{x}{16} = \frac{15}{40}$$

$$40x = 16 \times 15$$

$$\frac{40x}{40} = \frac{16 \times 15}{40}$$

$$x = \frac{16 \times 15}{40} = \frac{\cancel{2} \times \cancel{2} \times \cancel{2} \times 2 \times 3 \times \cancel{5}}{\cancel{2} \times \cancel{2} \times \cancel{2} \times \cancel{5}} = 6$$

It may look as if the techniques used in Chapter 6 give the answer more quickly, but there are advantages to the cross-multiplication technique, especially when our proportion is *not* a linear equation, as we will see in the next example.

Example 2

Solve the proportion: $\dfrac{3}{2x} = \dfrac{5}{7}$

Solution: This proportion is not a linear equation, because the variable is in the denominator. The power that the variable is raised is therefore –1 (because it is in the denominator; remember your rules for exponents: to move things from the denominator into the numerator you must change the sign of the exponent). Because this is not a linear equation, we will want to cross-multiply. Once we do that, we will have a linear equation that we can solve for x:

$$\frac{3}{2x} = \frac{5}{7}$$

Cross-multiply. $21 = 10x$

Interchange the two sides. $10x = 21$

Divide both sides by 10. $\dfrac{\cancel{10}x}{\cancel{10}} = \dfrac{21}{10}$

Simplify. Because the denominator is 10, writing the answer as a decimal number is natural. $x = 2.1$

Example 3

At the local print shop, 25 copies cost $1.50. At this rate, how much would 60 copies cost?

Solution: Set up a proportion. Let x represent that cost of 60 copies. Then 60 is to x as 25 is to $1.50. This can be translated into a proportion:

$$\frac{60}{x} = \frac{25}{1.5}$$

Now it is a matter of cross-multiplying and solving for x:

$$\frac{60}{x} = \frac{25}{1.5}$$

$$60 \times 1.5 = 25x$$

$$90 = 25x$$

$$25x = 90$$

$$\frac{\cancel{25}x}{\cancel{25}} = \frac{90}{25}$$

$$x = \frac{18}{5}$$

Money should almost always be given as a decimal. In order to get a decimal representation for this answer, we will need to divide:

$$
\begin{array}{r}
3.6 \\
5{\overline{\smash{\big)}\,18}} \\
\underline{15} \\
3.0 \\
\underline{3.0} \\
0
\end{array}
$$

So, 60 copies will cost \$3.60.

Example 4

One cubic centimeter of gold has a mass of 19.3 grams. A gold bar held in the Federal Reserve Bank is 727.9 cubic centimeters. Find the mass of a gold bar.

Solution: Let x represent the mass of a gold bar. Then 1 cubic centimeter is to 19.3 grams as 727.9 cubic centimeters is to x. Use this information to set up a proportion:

$$\frac{1}{19.3} = \frac{727.9}{x}$$

Now cross-multiply and solve for x.

$$\frac{1}{19.3} = \frac{727.9}{x}$$
$$1 \cdot x = 19.3 \times 727.9$$
$$x = 14{,}048.47$$

So, one gold bar has a mass of 14,048.47 grams.

Example 5

One pound corresponds to a mass of 454 grams. How much does a gold bar weigh?

Solution: Let x represent the weight of a gold bar. Set up a proportion for this problem. One pound is to 454 grams as x pounds is to 14,048.47 grams:

$$\frac{1}{454} = \frac{x}{14,048.47}$$

Cross-multiply and solve for x:

$$\frac{1}{454} = \frac{x}{14,048.47}$$

$$454x = 1 \times 14,048.47$$

$$\frac{454x}{454} = \frac{14,048.47}{454}$$

$$x = 30.94$$

So, one gold bar weighs 30.94 pounds.

Lesson 7-2 Review

Solve the following problems.

1. $\dfrac{2}{x} = \dfrac{5}{100}$

2. $\dfrac{x}{8} = \dfrac{9}{12}$

3. $\dfrac{x+3}{10} = \dfrac{3}{5}$

4. An online bookstore is offering electronic books at a rate of 10 pages for $0.75. A printed version of a book that is 160 pages sells for $9.95. If you plan to read the entire book, which would be cheaper: buying the electronic version or the printed version of the book?

Lesson 7-3: Proportions and Percents

A proportion is an equation involving the ratios of two quantities. A ratio consists of a numerator and a denominator. In general, the denominator of a ratio can be any non-zero number. **Percents** are special ratios that always have a denominator equal to 100. We consider 100% to be equivalent to a whole: 100% = 1. We can use a proportion to convert a ratio into a percent.

A typical percent problem involves a proportion such as $\frac{a}{b} = \frac{c}{100}$. The variable a represents **the part**, b represents **the whole**, and c represents **the percent**. A percent problem can require you to solve for *the part, the whole*, or *the percent*. The key will always be to set up a proportion. To solve the proportion, cross-multiply and solve for the unknown piece.

$$\frac{\text{part}}{\text{whole}} = \frac{\text{percent}}{100}$$

Example 1

Convert $\dfrac{37}{40}$ into a percent.

Solution: In this case, you are given the part (37) and the whole (40), and you are asked to find the percent. Set up a proportion and call the percent x:

$$\frac{37}{40} = \frac{x}{100}$$

To solve for x, cross-multiply and apply the techniques discussed earlier:

$$\frac{37}{40} = \frac{x}{100}$$

$$37 \times 100 = 40x$$

$$\frac{\cancel{40}x}{\cancel{40}} = \frac{37 \times 100}{40} = \frac{37 \times \cancel{2} \times \cancel{2} \times \cancel{5} \times 5}{\cancel{2} \times \cancel{2} \times 2 \times \cancel{5}} = \frac{37 \times 5}{2} = 92.5$$

So $\dfrac{37}{40}$ represents 92.5%.

Example 2

Find 65% of 360.

Solution: In this case you are given the whole (360) and the percent (65%), and you are asked to find the part. Set up a proportion and let x represent the part. Then cross-multiply and solve for x:

$$\frac{x}{360} = \frac{65}{100}$$

$$100x = 65 \times 360$$

$$\frac{\cancel{100}x}{\cancel{100}} = \frac{65 \times 360}{100} = \frac{23400}{100} = 234$$

So, 65% of 360 is 234.

Example 3

216 is 48% of what number?

Solution: In this case, you are given the part (216) and the percent (48%), and you are asked to find the whole. Set up a proportion and let x represent the whole:

$$\frac{216}{x} = \frac{48}{100}$$

$$216 \times 100 = 48x$$

$$\frac{\cancel{48}x}{\cancel{48}} = \frac{216 \times 100}{48} = \frac{\cancel{2} \times \cancel{2} \times \cancel{2} \times \cancel{3} \times 3 \times 3 \times \cancel{2} \times 2 \times 5 \times 5}{\cancel{2} \times \cancel{2} \times \cancel{2} \times \cancel{2} \times \cancel{3}} = 450$$

So, 216 is 48% of 450.

Percents can be greater than 100%. For example, when preparing for a math test I know that every one of my students gives 110%!

Example 4

Find 150% of 30.

Solution: You are given the percent (150%) and the whole (30) and you are asked to find the part. Let x represent the part and set up a proportion to solve for x:

$$\frac{x}{30} = \frac{150}{100}$$

$$100x = 30 \times 150$$

$$\frac{\cancel{100}x}{\cancel{100}} = \frac{30 \times 150}{100} = \frac{4500}{100} = 45$$

So, 150% of 30 is 45.

Lesson 7-3 Review

Solve the following problems.

1. Convert $\frac{8}{25}$ into a percent.

2. Find 30% of 2,520.

3. 420 is 20% of what number?

4. Find 225% of 20.

Lesson 7-4: Percents and Equations

One way to solve percent problems is to set up a proportion. You may have noticed that once you cross-multiply, you end up with a linear equation to solve. You can skip the proportion stage and jump right to the linear equation.

If you start with the proportion

$$\frac{a}{b} = \frac{c}{100}$$

and cross-multiply, you have the linear equation:

$$b \times c = a \times 100$$

Remember that a is the part, b is the whole, and c is the percent. This equation translates to:

The whole times the percent is equal to the part times 100.

If you divide both sides of the equation by 100, you can simplify the equation further:

$$a = b \times \frac{c}{100}$$

If you associate the 100 with the percent, you can convert the percent, c, to a decimal by moving the decimal point two places to the left. Remember that dividing a number by a positive power of 10 moves the decimal point to the left. You can also convert a decimal number to a percent by multiplying by 100, or moving the decimal point two places to the right and tacking on a % symbol. The % symbol means "out of 100." For example, 90% has a decimal representation 0.90, and 0.6% has a decimal representation of 0.006.

The more you work with percents, the easier it is to set up these equations, but if you ever forget how to set up this equation, reach for your proportions.

Example 1

What is 85% of 560?

Solution: You are given the percent (85%) and the whole (560), and you need to find the part. Convert 85% to a decimal form (to get 0.85) and set up the equation with x representing the part:

$$x = 560 \times 0.85 = 476$$

So, 85% of 560 is 476.

Example 2

What percent of 65 is 52?

Solution: You are given the part (52) and the whole (65), and are asked for the percent. Let x represent the percent (as a decimal) and set up an equation:

$$52 = 65 \cdot x$$

$$\frac{\cancel{65}x}{\cancel{65}} = \frac{52}{65}$$

$$x = 0.8$$

Convert your decimal to a percent by moving the decimal point two places to the right and adding a % symbol: 52 is 80% of 65.

Lesson 7-4 Review

Solve the following problems.

1. What is 55% of 1,900?

2. What percent of 75 is 60?

3. 72 is 60% of what number?

Lesson 7-5: Percent Change

The **change** in a variable is the difference between the final value of the variable and the initial value of the variable. For example, the price of a gallon of gasoline last week was $2.85, and this week the price is $2.52. The change of the price of a gallon of gasoline is $2.52 − $2.85, or $0.33. When the change of a variable is negative, it means that the variable has decreased. When the change of a variable is positive, it means that the variable has increased. We use the symbol Δ to represent the change of a variable:

$$\Delta x = x_{\text{final}} - x_{\text{initial}}$$

The **percent change** of a variable is the ratio of the change in the variable divided by the initial value of the variable. This answer must be written as a percent, so you will need to convert the fraction to a percent. If the variable increases, then we talk about the **percent increase** of the variable. If the variable decreases, we talk about the **percent decrease**. Both the percent increase and the percent decrease are positive numbers. The sign of the percent change is included through the words *increase* and *decrease*.

Example 1

Find the percent increase from 40 to 45.

Solution: The percent change is found by first evaluating the ratio:

$$\frac{\text{final-initial}}{\text{initial}}$$

$$\frac{\text{final} - \text{initial}}{\text{initial}} = \frac{45 - 40}{40} = \frac{5}{40}$$

Convert $\frac{5}{40}$ to a percent:

$$\frac{5}{40} = \frac{x}{100}$$

$$500 = 40x$$

$$\frac{\cancel{40}x}{\cancel{40}} = \frac{500}{40} = \frac{5 \times \cancel{2} \times \cancel{2} \times \cancel{5} \times 5}{\cancel{2} \times \cancel{2} \times 2 \times \cancel{5}} = \frac{25}{2} = 12.5$$

So, the percent increase is 12.5%.

As an alternative, you could have first converted $\frac{5}{40}$ to a decimal and then changed the decimal representation to a percent by moving the decimal point two places to the right: $\frac{5}{40} = 0.125$, or 12.5%.

Example 2

Find the percent decrease from 42 to 35.

Solution: The first step in finding the percent decrease is to find the percent change. The percent change is found by first evaluating the ratio:

$$\frac{\text{final-initial}}{\text{initial}}$$

$$\frac{\text{final-initial}}{\text{initial}} = \frac{35 - 42}{42} = \frac{-7}{42} = -\frac{1}{6}$$

Convert $-\frac{1}{6}$ to a decimal number by dividing: $-\frac{1}{6} = -0.167$.

Convert the decimal representation to a percent by moving the decimal point two places to the right: $-0.167 = -16.7\%$. The percent *change* is -16.7%, and the percent *decrease* is 16.7%.

You don't usually talk about a negative percent decrease. That is similar to using two negatives in a sentence. Technically, two negatives make a positive, but sometimes two negatives is just confusing. So choose your terminology appropriately: Either the percent decrease is a positive 16.7% (with the word *decrease* indicating the direction of the change) or the percent change is –16.7%.

Example 3

The population of Mishicot, Wisconsin, has grown from 2,000 to 2,200. Find the percent change and the percent increase in the population.

Solution: The first step in finding the percent increase is to find the percent change. The percent change is found by first evaluating the ratio:

$$\frac{\text{final-initial}}{\text{initial}}$$

$$\frac{\text{final-initial}}{\text{initial}} = \frac{2200 - 2000}{2000} = \frac{200}{2000} = \frac{1}{10}$$

Next, find the decimal equivalent of $\frac{1}{10}$ and then convert it to a percent by moving the decimal point two places to the right:

$$\frac{1}{10} = 0.1 = 10\%$$

The percent *change* in the population is 10%, and the percent *increase* in the population is 10%.

Example 4

The weight of the space shuttle before launch is 2,041,166 kilograms. At the end of the mission, the space shuttle weighs 104,326 kilograms. Find the percent change and the percent decrease of the weight of the space shuttle.

Solution: The percent change is found by evaluating the ratio:

$$\frac{\text{final-initial}}{\text{initial}}$$

$$\frac{\text{final-initial}}{\text{initial}} = \frac{104,326 - 2,041,166}{2,041,166} = \frac{-1,936,840}{2,041,166} = -0.95$$

The percent change is –95%, and the percent decrease is 95%. The space shuttle loses around 95% of its weight over the course of a mission.

Lesson 7-5 Review

Solve the following problems.

1.   Find the percent increase from 50 to 65.

2.   The price of dog food has gone from $30 per bag to $33 per bag. Find the percent increase.

3.   Switching cell phone plans has resulted in a decrease in my monthly phone bill from $40 to $35. Find the percent change and the percent decrease in my monthly phone bill.

Lesson 7-6: Mark-up and Discount

Stores make a profit by buying their products at a lower price than they sell them. The amount that a store increases the price is called the **mark-up**:

$$\text{mark-up} = \text{price}_{\text{consumer}} - \text{price}_{\text{store}}$$

The mark-up is the difference between what the store charges the consumer and what the store actually pays for the item. The percent change in the price is called the **percent mark-up**. The percent mark-up can be found by dividing the mark-up by the store's price:

$$\text{percent mark-up} = \frac{\text{mark-up}}{\text{store's price}}$$

The percent mark-up can also be found using the ratio

$$\text{percent mark-up} = \frac{\text{price}_{\text{consumer}} - \text{price}_{\text{store}}}{\text{price}_{\text{store}}}$$

If we are given the percent mark-up and one of prices, either the price that the store *pays* or the price that the store *charges* the consumer, we can use this ratio to determine the other price. If we are given both prices, we can first find the mark-up and then find the percent mark-up. Make sure that you convert the percent mark-up to its decimal equivalent when working out these calculations.

Example 1

A grocery store has a 70% mark-up on its beef. If the store pays $9 per pound for filet mignon, find the mark-up and the price to the consumer.

Solution: Convert the percent mark-up to a decimal, and use the equation for percent mark-up to determine the mark-up:

$$\text{percent mark-up} = \frac{\text{mark-up}}{\text{store's price}}$$

$$0.7 = \frac{\text{mark-up}}{9}$$

$$9 \times 0.7 = 9 \times \frac{\text{mark-up}}{9}$$

$$\text{mark-up} = 6.3$$

The mark-up is $6.30 per pound. The price to the consumer is the store's cost plus the mark-up. The price to the consumer is therefore $9 + $6.30, or $15.30.

Example 2

A computer store has a 75% mark-up. If it pays $120 for an all-in-one printer, find the selling price of the printer.

Solution: Use the equation for the percent mark-up to determine the mark-up:

$$\text{percent mark-up} = \frac{\text{mark-up}}{\text{store's price}}$$

Let m represent the mark-up:

$$0.75 = \frac{m}{120}$$

$$120 \times 0.75 = 120 \times \frac{m}{120}$$

$$m = 90$$

Now that we know the mark-up, we can find the selling price, or the price to the consumer:

$$\text{mark-up} = \text{price}_{consumer} - \text{price}_{store}$$

$$\text{price}_{store} + \text{mark-up} = \text{price}_{consumer}$$

$$120 + 90 = \text{price}_{consumer}$$

$$\text{price}_{consumer} = 210$$

The selling price is $210.

When an item goes on sale, the selling price is reduced. The amount of the price decrease is called the discount, or mark-down. The equation that models this situation is:

$$\text{price}_{sale} = \text{price}_{regular} - \text{discount}$$

The percent discount is the ratio of the discount to the regular price. It can be calculated using the equation:

$$\text{percent discount} = \frac{\text{discount}}{\text{regular price}}$$

Example 3

A jacket that usually sells for $60 is on sale for 20% off. Find the discount.

Solution: Convert the percent discount to a decimal and use the equation for percent discount to find the discount:

$$\text{percent discount} = \frac{\text{discount}}{\text{regular price}}$$

$$0.2 = \frac{\text{discount}}{60}$$

$$60 \times 0.2 = 60 \times \frac{\text{discount}}{60}$$

$$\text{discount} = 12$$

The discount is $12.

Example 4

A computer that regularly sells for $600 is on sale for 30% off. Find the sale price of the computer.

Solution: First find the discount, and then find the sale price:

$$\text{percent discount} = \frac{\text{discount}}{\text{regular price}}$$

$$0.3 = \frac{\text{discount}}{600}$$

$$600 \times 0.3 = 600 \times \frac{\text{discount}}{600}$$

$$\text{discount} = 180$$

Now find the sale price:

$$\text{price}_{\text{sale}} = \text{price}_{\text{regular}} - \text{discount}$$

$$\text{price}_{\text{sale}} = 600 - 180$$

$$\text{price}_{\text{sale}} = 420$$

The sale price is $420.

There is a common misconception about the relationship between mark-up and discount that I will now address. Suppose that a clothing store has a 55% mark-up on business suits and that the store pays $100 for each business suit. The store will mark-up the business suit by $55, and charge $155. Now, suppose that the business suits are not selling very well, and the store decides to discount the suits by 55%. In other words, the business suits were initially marked up 55% and then they

7 RATIOS, PROPORTIONS, AND PERCENTS

were discounted 55%. Let's compare the sale price to the price that the store paid for each business suit.

The mark-up percent was 55%, so the price to the consumer was $155. The discount percent is also 55%, so the discount will be 0.55 × 155, which is $85.25. That means that the sale price will be $155 – $85.25, or $69.75. That's quite a bit lower than the store price! This result may go against your intuition. It would seem that if you mark-up and then discount by the same percent, then you would be back where you started, but this is not the case. The reason for this is that your mark-up percent is applied to the lower price, and the discount percent is applied to the higher price. If the mark-up percent and the discount percent are the same value, then the mark-up amount will be less than the discount amount. You will end up at a lower price than where you started.

Lesson 7-6 Review

Solve the following problems.

1. A bookstore has a 40% mark-up. If the store pays $6 for a book, find the mark-up.

2. A bookstore has a 40% mark-up. If the store pays $6 for a book, find the selling price.

3. A bookstore has a 40% mark-up. If the selling price of a book is $42, find the price that the bookstore pays for the book.

4. A shoe store is having a sale on Doc Szecsei boots. The regular price is $120 and the percent discount is 15%. Find the sale price.

5. An electronics company is having a sale on iPort MP3 players. If the regular price of the iPort is $150 and the percent discount is 20%, find the sale price.

Answer Key

Lesson 7-1 Review

1. A 5-pound bag costs $1.70 per pound, a 20-pound bag costs $1.07 per pound, and the 40-pound bag costs $0.75 per pound.

2. A 12-pack costs $0.25 per can, a 24-pack costs $0.21 per can.

3. The car gets 25 miles per gallon in the city and 30 miles per gallon on the interstate.

4. Betty drives at a rate of 70 miles per hour.

Lesson 7-2 Review

1. $x = 40$

2. $x = 6$

3. $x = 3$

4. Let x represent the price for the online book and set up a proportion: $\frac{0.75}{10} = \frac{x}{160}$. Solve for x: $x = 12$. The online book would cost $12, and the printed version would cost $9.95. The printed version would cost less.

Lesson 7-3 Review

1. 32%

2. 756

3. 2,100

4. 45

Lesson 7-4 Review

1. 1,045

2. 80%

3. 120

Lesson 7-5 Review

1. 30%

2. 10%

3. percent change: –12.5%; percent decrease: 12.5%

7
RATIOS, PROPORTIONS, AND PERCENTS

Lesson 7-6 Review

1. $2.40

2. $8.40

3. $30

4. $102

5. $120

8

Inequalities and Graphs

Solving equations involves working with equalities. We have some guidelines for how to transform equations so that the solution can be determined easily. These transformations are possible because *equality* is an equivalence relation. Remember that equality is an example of an equivalence relation, which is a relation that has the reflexive, symmetric, and transitive properties. It's time to introduce you to two new relations: less than and greater than. We call these two relations *inequalities*.

Lesson 8-1: Inequalities

An **inequality** is an algebraic statement that compares two algebraic expressions that are not necessarily equal. I mentioned four basic inequalities in Chapter 1 and I'll summarize them here:

Symbol	Meaning	Example
<	Is less than	$5 < 8$
≤	Is less than or equal to	$5 \le 8$, $5 \le 5$
>	Is greater than	$8 > 5$
≥	Is greater than or equal to	$8 \ge 5$, $5 \ge 5$

A solution to an inequality is a number that produces a true statement when it is substituted for the variable in the inequality. Inequalities are more flexible than equalities and are therefore easier to satisfy than are equalities. There is only one number that is equal to 3, even though there are many ways to write the number 3. We can write 3 as $4 - 1$, or $2 + 1$, or $2 \times \frac{3}{2}$, but no matter how you slice it, they all represent the same number. With inequalities, there are a whole lot of numbers that are greater than 3. I can think of several numbers right away, such as 4, 5, 6, 7, and 8, and I am just getting started. These numbers are all different from each other; they are not different representations of the same number. Because of the flexibility in satisfying an inequality, inequalities usually have many solutions.

We can examine which of these inequalities, if any, are equivalence relations. I will examine $<$ and $\leq$ in more detail.

To determine whether $<$ (less than) is an equivalence relation, we need to check to see if it has all three required properties:

1. Reflexive property: Is $a < a$? Is a number less than itself? No.

2. Symmetric property: If $a < b$ is $b < a$? No.

3. Transitive property: If $a < b$ and $b < c$, is $a < c$? Yes.

So, $<$ is not an equivalence relation because it only has the transitive property; it does not have the reflexive or symmetric property.

To determine whether $\leq$ (less than or equal to) is an equivalence relation, we need to check to see if it has all three required properties:

1. Reflexive property: Is $a \leq a$? Yes.

2. Symmetric property: If $a \leq b$ is $b \leq a$? Not always, so no.

3. Transitive property: If $a \leq b$ and $b \leq c$, is $a \leq c$? Yes.

So, $\leq$ is not an equivalence relation because it only has the reflexive and transitive properties; it does not have the symmetric property.

You may be wondering why I spent so much time talking about equivalence relations. In order to solve inequalities, we will need to establish some rules about how we are allowed to transform them. Inequalities are more flexible than equalities, and as a result, there are some transformations that we are *not* allowed to do with equalities that we *are* allowed to do with inequalities. The differences in how we treat equalities and inequalities stems from the fact that equality is an equivalence relation, whereas inequalities are not.

Lesson 8-2: Properties of Inequalities

There are two ways that we can transform equations: We can add (or subtract) the same number to both sides of the equality, and we can multiply (or divide) both sides of the equation by the same number. These transformations are allowed because of the addition property of equality and the multiplication property of equality. Inequalities actually have two addition properties and a restricted multiplication property.

The **first addition property of inequality** states that if $a > b$, then $a + c > b + c$. To understand this property, let's go back to our example with money. Suppose I have \$25 and you have \$5. If someone gives us each \$5, then I will still have more money than you. I will have \$30 and you will have \$10. We will both be richer, but I'll still be richer than you.

The **second addition property of inequality** states that if $a > b$ and $c > d$, then $a + c > b + d$. Again, this is not surprising. Suppose I have \$25 and you have \$5. If some kind person gives me \$10 and you \$5, then I will definitely still have more money than you.

The only thing you are not allowed to do is try to correct the inequality by trying to balance things out. Either you give the same amount to both sides of the inequality (the first addition property) or you increase the imbalance by giving more to the "haves" you do to the "have nots" (the second addition property). You are not allowed to try to make up for any injustice by giving more to the "have nots." It may not seem fair, but that's the way it is. That's why they are called *inequalities*.

The restricted **multiplication property of inequalities** is also fairly clear. It states that if $a > b$ and $c > 0$, then $a \cdot c > b \cdot c$. The restriction that the number that you multiply by must be positive is very important, and it is worth looking at in more detail.

Consider the inequality $0 < 1$. Suppose that we wanted to multiply both sides of this inequality by a negative number; let's multiply both sides of the inequality $0 < 1$ by –1. On the left hand side of the inequality, you would have $0 \cdot (-1)$, which is 0. On the right hand side of the inequality, you would have $1 \cdot (-1)$, which is –1. Now, how does 0 compare to –1? Well, $0 > -1$. Notice that we started with $0 < 1$, and when we multiplied both sides of the inequality $0 < 1$ by –1, we end up with the inequality $0 > -1$. This leads us to an important rule about multiplying an inequality by a negative number: When you multiply an inequality by a negative number, you must flip the inequality symbol. We can write this mathematically as:

$$\text{If } a < b \text{ and } c < 0 \text{ then } a \cdot c > b \cdot c.$$

This flipping rule also holds when you divide both sides of an inequality by a negative number. Keep in mind that the focus is on the *sign* of the number by which you are multiplying both sides of the inequality.

Lesson 8-3: Solving Inequalities in 1 Step

Solving an inequality is very similar to solving an equality. The goal is still to isolate the variable on one side of the inequality. The transformation rules for inequalities are similar to those for equality. One of the main differences in the procedure occurs when you multiply or divide both sides of an inequality. You must be sure to flip the inequality when you multiply (or divide) both sides by a negative number.

The other difference between solving equalities and solving inequalities is that with equalities, if you have an equation such as $3 = x$, you are allowed to interchange the sides and write $x = 3$. The reason you can do this is because equality has the symmetric property. Inequalities do not have the symmetric property, so if you have an inequality such as $3 < x$ and you want to turn the inequality around, you must write $x > 3$. If you interchange the sides of an inequality, the inequality symbol must be turned around so that it "points" to the same expression that it pointed to originally.

The addition transformations are summarized in this table:

Transformation	Original Equation	Transformation	New Equation
Simplify one or both sides	$x > 6 + 9$	Simplify	$x > 15$
Add the same number to each side and simplify	$x - 5 > 9$	Add 5 to both sides and simplify	$x - 5 + 5 > 9 + 5$ $x > 14$
Subtract the same number from each side and simplify	$x + 7 > 12$	Subtract 7 from both sides and simplify	$x + 7 - 7 > 12 - 7$ $x > 5$
Turn the inequality around	$3 < x$	Turn the entire inequality around	$x > 3$

Example 1

Solve the inequality: $x + 3 < 8$.

Solution: Subtract 3 from both sides and simplify:

$x + 3 < 8$

$x + 3 - 3 < 8 - 3$

$x < 5$

The solution is all real numbers less than 5.

Example 2

Solve the inequality: $-10 > x - 6$

Solution: Add 6 to both sides and simplify:

$-10 > x - 6$

$-10 + 6 > x - 6 + 6$

$-4 > x$

$x < -4$

The solution is all real numbers less than -4.

The transformations involving multiplication and division are summarized in the following table: (Pay special attention to the transformations that involve multiplying or dividing by a negative number.)

Transformation	Original Equation	Transformation	New Equation
Multiply both sides by the same *positive* number and simplify	$\frac{1}{2}x > 9$	Multiply both sides by 2 and simplify	$2 \cdot \left(\frac{1}{2}x\right) > 2 \cdot 9$ $x > 18$
Divide both sides by the same *positive* number and simplify	$2x > 8$	Divide both sides by 2 and simplify	$\frac{2x}{2} > \frac{8}{2}$ $x > 4$
Multiply both sides by the same *negative* number, reverse the inequality, and simplify	$-\frac{2}{3}x > 4$	Multiply both sides by $-\frac{3}{2}$, flip the inequality symbol, and simplify	$\left(-\frac{3}{2}\right)\left(-\frac{2}{3}x\right) < \left(-\frac{3}{2}\right)4$ $x < -6$
Divide both sides by the same *negative* number, reverse the inequality, and simplify	$-3x > 4$	Divide both sides by -3, flip the inequality symbol, and simplify	$\frac{-3x}{-3} < \frac{4}{-3}$ $x < -\frac{4}{3}$

Example 3

Solve the inequality: $\dfrac{6}{11}x > 3$

Solution: Multiply both sides by $\dfrac{11}{6}$ and simplify:

$$\frac{6}{11}x > 3$$

$$\frac{11}{6} \cdot \frac{6}{11}x > \frac{11}{6} \cdot 3$$

$$x > \frac{11}{2}$$

The solution is all real numbers greater than $\dfrac{11}{2}$.

Example 4

Solve the inequality: $4x > 10$

Solution: Divide both sides by 4 and simplify:

$$4x > 10$$

$$\frac{4x}{4} > \frac{10}{4}$$

$$x > \frac{5}{2}$$

The solution is all real numbers greater than $\dfrac{5}{2}$.

Example 5

Solve the inequality: $-3x < 15$

Solution: Divide both sides by –3, flip the inequality symbol, and simplify:

$$-3x < 15$$

$$\frac{-3x}{-3} > \frac{15}{-3}$$

$$x > -5$$

The solution is all real numbers greater than –5.

8

INEQUALITIES AND GRAPHS

Example 6

Solve the inequality: $-\dfrac{1}{3}x > -5$

Solution: Multiply both sides by –3, flip the inequality symbol, and simplify:

$$-\frac{1}{3}x > -5$$

$$(-3)\left(-\frac{1}{3}x\right) < (-3)(-5)$$

$$x < 15$$

The solution is all real numbers less than 15.

Lesson 8-3 Review

Solve the following inequalities.

1. $x + 5 < 12$
2. $-2 > x - 5$
3. $\dfrac{4}{15}x > 8$
4. $3x > 5$
5. $-4x < 20$
6. $-\dfrac{1}{2}x > -4$

Lesson 8-4: Solving Inequalities in 2 Steps

In general, when you want to solve an inequality involving one variable, the first thing you need to do is move all of the terms that involve the variable over to one side of the inequality and move all of the terms that don't involve the variable over to the other side. If, after combining all of the terms together, the coefficient in front of the variable is a number other than 1, you will need to multiply both sides of the inequality by the reciprocal of the coefficient in front of the variable. Hopefully this process sounds familiar. It is similar to the method we used to solve equalities using two steps.

Example 1

Solve the inequality: $\frac{1}{3}x + 4 > -8$

Solution: Subtract 4 from both sides, then multiply by 3, and simplify:

$$\frac{1}{3}x + 4 > -8$$

Subtract 4 from both sides.

$$\frac{1}{3}x + 4 - 4 > -8 - 4$$

Simplify.

$$\frac{1}{3}x > -12$$

Multiply both sides by 3.

$$3 \cdot \left(\frac{1}{3}x\right) > 3 \cdot (-12)$$

Simplify.

$$x > -36$$

The solution is all real numbers greater than –33.

Example 2

Solve the inequality: $2x + 10 \geq 7x$

Solution: Collect all the terms involving variables on one side, and the numbers on the other, and then solve for x:

$$2x + 10 \geq 7x$$

Subtract 2x from both sides.

$$2x + 10 - 2x \geq 7x - 2x$$

Simplify.

$$10 \geq 5x$$

Divide both sides by 5.

$$\frac{10}{5} \geq \frac{5x}{5}$$

Simplify.

$$2 \geq x$$

Turn the entire inequality around, if you like seeing the variable on the left (as I do).

$$x \leq 2$$

The solution is all real numbers less than or equal to 2.

Example 3

Solve the inequality: $12 > -2x - 6$

Solution: Add 6 to both sides, and then divide both sides by –2. Remember to flip the inequality symbol:

$$12 > -2x - 6$$

Add 6 to both sides.

$$12 + 6 > -2x - 6 + 6$$

Simplify.

$$18 > -2x$$

Divide both sides by –2 and flip the inequality.

$$\frac{18}{-2} < \frac{-2x}{-2}$$

Simplify.

$$-9 < x$$

Turn the entire inequality around, if you like seeing the variables on the left (as I do).

$$x > -9$$

The solution is all real numbers greater than –9.

Lesson 8-4 Review

Solve the following inequalities.

1. $\frac{1}{5}x + 1 < 5$

2. $3 < -4x - 8$

3. $3x > 5x - 8$

Lesson 8-5: Graphing Inequalities Using the Number Line

The solutions to the inequalities in the last section were represented algebraically. The answer was not unique; there were many real numbers that satisfied the inequalities. We described the collection of these solutions with words, but a picture is worth a thousand words. It helps to understand the solution to an inequality by being able to visualize it. To do this, we will make use of the number line.

Zero is one of the most important points on a number line. The point 0 is called the origin. Negative numbers lie to the left of the origin, and positive numbers lie to the right. To draw a number line, draw a line with arrows pointing in both directions, and label a point 0. It's best to keep your number lines simple. Some people start labeling lots of integers around 0, but I recommend only labeling points that are pertinent. And at this point, the only important point is 0. Take a look at my number line shown in Figure 8.1.

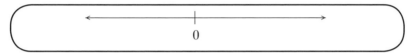

0

Figure 8.1.

A ray looks as if it's a line that is cut in half. A ray has only one arrow; the other arrow has been cut off. It has a definite starting point, but it never ends. Any number on the number line can serve as a starting point for a ray. A ray either includes all of the numbers to the left of (or less than) the starting point, or all of the numbers to the right of (or greater than) the starting point. The set of all numbers less than a given number is drawn as an arrow starting at the given number and pointing to the left. The set of all numbers greater than a given number is drawn as an arrow starting at the given number and pointing to the right.

The starting point of a ray may or may not be included. The relations $<$ and $>$ are called **strict inequalities.** For strict inequalities, the starting point is *not* included in the ray. We draw an open circle to reflect that the starting point is not included. The relations $\leq$ and $\geq$ are just called inequalities, and for these inequalities, the starting point is included in the ray. We draw a closed circle to reflect that the starting point is included. A ray that does not include its starting point is called an **open ray**; a ray that does include its starting point is called a **closed ray.**

The following will help you draw a ray:

⊃   Start with a number line with 0 labeled.

8

INEQUALITIES AND GRAPHS

○ Label your starting point. If your starting point is negative, make sure you move to the left of 0. If the starting point is positive, move to the right of 0.

○ Look at the type of inequality involved to determine if your ray is open or closed.

⇨ If the inequality is *strict*, then the ray is open; draw an open circle at your labeled starting point.

⇨ If the inequality is *not strict*, then the ray is closed; draw a filled-in circle at your labeled starting point.

○ Draw a ray pointing in the correct direction from your starting point.

Figure 8.2 shows the open ray representing the set of real numbers less than 3.

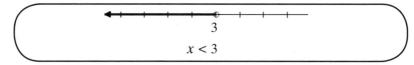

$x < 3$

Figure 8.2.

Figure 8.3 shows an open ray representing the set of real numbers *less than –3*, and the closed ray representing the set of real numbers *less than or equal to –3*. Notice that the only difference between these two rays is the circle at –3. The open ray has an open circle and the closed ray has a closed (or filled in) circle. Both of these rays point to the left because we are describing points that are less than (or less than or equal to) –3.

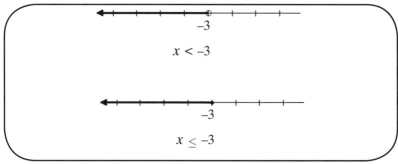

$x < -3$

$x \leq -3$

Figure 8.3.

Figure 8.4 shows an open ray representing the set of real numbers *greater than* 2, and the closed ray representing the set of real numbers *greater than or equal to* 2. Notice that the only difference between these two rays is the circle at 2. The open ray has an open circle and the closed ray has a closed (or filled in) circle. Both of these rays point to the right because we are describing points that are greater than (or greater than or equal to) 2.

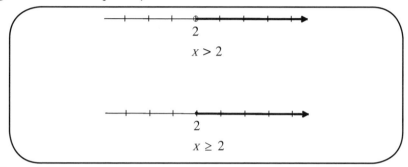

Figure 8.4.

In the last section, we solved several inequalities and described the solution in words. We will take a minute to give a graphical solution to those same inequalities.

Example 1

Graph the solution to the inequality: $-3x < 15$

Solution: The solution to this inequality is $x > -5$, which is the open ray shown in Figure 8.5.

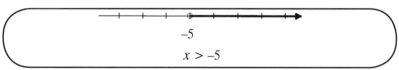

Figure 8.5.

Example 2

Graph the solution to the inequality: $2x + 10 \geq 7x$

Solution: The solution to this inequality is $x \leq 2$, which is the closed ray shown in Figure 8.6.

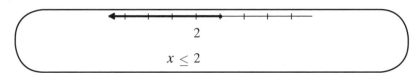

2

$x \leq 2$

Figure 8.6.

Example 3

Graph the solution to the inequality: $12 > -2x - 6$

Solution: The solution to this inequality is $x > -9$, which is the open ray shown in Figure 8.7.

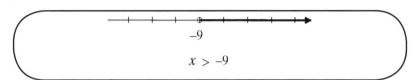

-9

$x > -9$

Figure 8.7.

Lesson 8-5 Review

Graph the solutions to the following inequalities.

1. $\frac{1}{5}x + 1 < 5$

2. $3 < -4x - 8$

3. $3x > 5x - 8$

Answer Key
Lesson 8-3 Review
1. $x < 7$

2. $x < 3$

3. $x < 30$

4. $x > \frac{5}{3}$

5. $x > -5$

6. $x < 8$

Lesson 8-4 Review
1. $x < 20$

2. $x < -\frac{11}{4}$

3. $x < 4$

Lesson 8-5 Review
1. Graph of $x < 20$

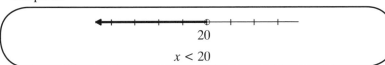

$$x < 20$$

2. Graph of $x < -\frac{11}{4}$

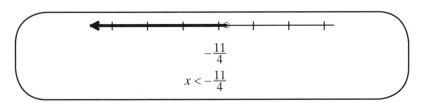

$$x < -\frac{11}{4}$$

3. Graph of $x < 4$

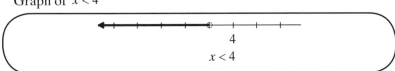

$$x < 4$$

9

Linear Functions and Graphs

Algebraic expressions can be thought of as a collection of instructions. The steps involved in evaluating the expression $2x + 4$ are:

⊃ Take the value of x and double it.

⊃ Take the number obtained in the previous step and add 4 to it.

When we talk about an algebraic expression, it is convenient to have a name to associate with it. We can call an algebraic expression anything we want. Most of the time we call algebraic expressions either $f(x)$ or y. The function name $f(x)$ is pronounced "f of x." The function $f(x)$ defined by $f(x) = 2x + 4$ would be read as "f of x equals two x plus four." The name is not as important as the set of instructions that goes along with it.

Once you name an algebraic expression, a function is born. A *function* usually involves two variables. One variable is contained in the algebraic expression and is called the **independent variable**. The other variable is the *name* of the function and is called the **dependent variable**. We think of the dependent variable *depending* on the values of the *independent* variable.

Before I talk about *linear* functions, I will need to discuss relations and functions in general, and point out their differences.

Lesson 9-1: Relations and Functions

A **relation** is a mathematical way of establishing a *relationship* between two quantities. One of the quantities in the relationship is called the **input**, and the other quantity is called the **output**. The input is also called the **domain**, and the output is often referred to as the **range** or the **co-domain**. Because a relation involves a relationship between *two* different quantities, we will need *two* variables to keep track of, or represent, these two quantities. The variable used to represent the input is called the *independent variable*, and the variable used to represent the output is called the *dependent variable*.

A relation is often written as a collection of ordered pairs. Ordered pairs are an efficient way to keep track of the relationship between the input values and the output values. An **ordered pair** consists of two elements, the first element coming from the domain and the second one coming from the range. The two elements are written side by side, separated by a comma and surrounded by parentheses. I have just used a lot of words to describe an object that looks this way: (a, b), where a is an object in the domain and b is an object in the range that is related to a. If the independent variable is called x, we call a the **x-coordinate** of the ordered pair, and if the dependent variable is called y, we call b the **y-coordinate** of the ordered pair.

Relations are fairly relaxed in the sense that there are no restrictions or regulations regarding the number of output values that can be associated with a particular input value. The input values are values that the independent variable takes on, and the output values are the corresponding values for the dependent variable.

Relations can be described in a variety of ways. You can describe a relation using words, a formula, a table of input and output values, a list of ordered pairs, or a graph.

It may be helpful to give you a couple of examples of relations. Here's a list of ordered pairs: (1, 3), (2, 6), (2, 4), (3, 9). This same relation could be represented using the following table.

Input	Output
1	3
2	6
2	4
3	9

The problem with tables and lists of ordered pairs is that you are limited to the data provided. You would not be able to determine the output value that corresponds to an input value of 4 for this relation. Not only that, but the output isn't necessarily a unique value determined by the input. As you can see in the previous table, an input of 2 results in two different outputs: 6 and 4. This is one of the problems associated with relations.

Relations can also be specified by writing a formula and using variables to represent the input and the output. A formula gives a set of specific instructions about what to do with the input value. A **formula** is an equation where one side of the equal sign is the dependent variable and the other side of the equal sign consists of an algebraic expression that only involves the independent variable. For example, if x represents the input of our relation, or independent variable, and y represents the output, or the dependent variable, of our relation, the formula $y = 2x + 1$ specifies a relation between x and y. You could use this formula to calculate the output for various values of the input. Having a formula enables you to completely describe a relation without having to list every single ordered pair in the relation.

Relations that are described using words are very important in applying mathematics to our everyday lives. For example, if your cellular phone plan costs $35 per month for up to 300 minutes and 39 cents for every minute over 300, you can use this relationship between the minutes that you use in a month and your expected phone bill for the month.

Relations can be described graphically. The graph of a relation helps you visualize that relation so that you can understand its properties more thoroughly. For example, you could look at a graph of the price of a particular stock over time and decide whether to buy more shares of that stock.

Just like a relation, a **function** is a rule that establishes a relationship between two quantities and has an input and an output. The input is still called the domain, and the output is still called the range or the co-domain. The variable used to describe the elements in the domain is called the independent variable, and the variable used to describe the elements in the range is called the dependent variable. But there's more to a function than there is to a relation.

With a relation, the same input value could result in two different output values. With a function, this is not allowed. The feature that distinguishes a function from a relation is that with a function, each input value is assigned a *unique* output value. We tend to focus on functions more than relations because of this extra stipulation.

It is easy to become attached to algebraic expressions and give them names. I told you in the beginning of this chapter that as soon as you name an algebraic expression, a function is born. Unfortunately, mathematicians aren't very clever with their names; y and $f(x)$ are the most common. Keep in mind that it doesn't matter what we name an expression, just as it doesn't matter what we name a variable. We will interchange y and $f(x)$ frequently, sometimes in the middle of working out one problem! It's important that you remember that y and $f(x)$ are common names to denote an expression.

A function is a specific type of relation, and, because of that, we can represent a function in the same way that we can represent a relation: words, a formula, a table of input and output values, a list of ordered pairs, or a graph to describe a function.

Lesson 9-2: Linear Functions and Equations

A linear function is one of the nicest functions to work with. A **linear function** is created when we name a linear *expression*. Remember that a linear expression contains only constants and one variable raised to the first power. The function $f(x) = 3x + 2$ is an example of a linear function. If the function is called y instead of $f(x)$ you will see it written as $y = 3x + 2$. Don't let the names bother you; it doesn't matter how we name the function. What matters is the formula that describes it. Two functions that have different names, but are described using the same formula, represent the same function.

Linear functions are usually written as $f(x) = mx + b$, where m and b are constants and x is the independent variable. The constants m and b are very important and have special names. The constant m is called the slope, and the constant b is called the y-intercept. Functions of this form are called linear because their graphs are lines. Every linear function involves two constants, the slope and the y-intercept, and one independent variable with an exponent equal to 1.

We can evaluate functions for particular values of the independent variable.

Example 1

Evaluate the function $y = 3x + 2$ when $x = 2$ and write your answer as an ordered pair.

Solution: When $x = 2$, $y = 3 \cdot 2 + 2 = 6 + 2 = 8$. This corresponds to the ordered pair (2, 8).

Example 2

Evaluate the function $y = -3x + 5$ when $x = -1$ and write your answer as an ordered pair.

Solution: When $x = -1$, $y = (-3)(-1) + 5 = 3 + 5 = 8$. This corresponds to the ordered pair (–1, 8).

Remember that functions are special collections of ordered pairs. When a function is defined using a formula, we are actually writing an *equation*, with the dependent variable on one side of the equality and

9

LINEAR FUNCTIONS AND GRAPHS

an *algebraic expression* involving the independent variable on the other side. The ordered pairs that make up the function are generated by evaluating the algebraic expression for various values of the independent variable. Each ordered pair that is generated using the equation that defines the function is called a **solution to the equation**, or a point that satisfies the equation.

For a given linear function, there are many different values that the independent variable can take on, and for each value of the independent variable there is a unique corresponding value for the dependent variable. As a result, a linear equation can have many solutions, and it is helpful to draw a picture of all of the solutions of a particular equation. In order to draw such a picture, we need to talk about the Cartesian coordinate system.

Lesson 9-2 Review

Evaluate the following functions at the indicated value and write your answer as an ordered pair.

1. $f(x) = 5x - 1$ at $x = -1$
2. $f(x) = 2x$ at $x = 3$
3. $f(x) = -4x + 2$ at $x = -1$
4. $f(x) = 6x - 9$ at $x = 0$

Lesson 9-3: The Cartesian Coordinate System

To graph a function, you need to be able to graph both the input and the output at the same time. This amounts to being able to graph an ordered pair (a, b). We have already talked about how the number line can be used to graph a number. In order to graph an ordered pair (a, b), we will need to be able to graph two numbers simultaneously.

To graph an ordered pair, you will need to use one number line for the input and another number line for the output. If you arrange these two number lines so that they are perpendicular to each other, as is shown in Figure 9.1, you will have created what is known as the **Cartesian coordinate system.** The Cartesian coordinate system is named for Rene Descartes, who is credited with inventing it. This coordinate system brings geometry and algebra together, enabling us to use algebra to solve problems in geometry, and to use geometry to gain insight into algebraic results. We use this system, which is sometimes called the coordinate plane, to locate points and draw figures.

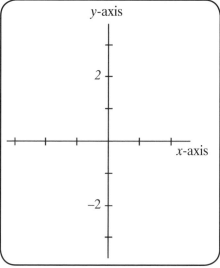

The horizontal number line is called the **x-axis**, and it is used to record the values of the input, or independent variable, or the first number in the ordered pair *Figure 9.1.* (*a*, *b*). The vertical number line is called the **y-axis**, and it is used to record the function values, or the output, or the values of the dependent variable, or the second number in the ordered pair (*a*, *b*). We will draw the two number lines so that they intersect at 0; the point where the two lines intersect is called the **origin.** The ordered pair corresponding to the origin is (0, 0).

Two numbers are used to describe the location of a point in the plane, and they are recorded as an ordered pair (*x*, *y*), where the first number represents the horizontal distance from the *y*-axis to the point and the second number represents the vertical distance from the *x*-axis to the point. The first coordinate is called the *x*-coordinate, and the second coordinate is called the *y*-coordinate.

Points that are on the *x*-axis have a *y*-coordinate equal to 0, and points that are on the *y*-axis have their *x*-coordinate equal to 0. Those points that lie to the right of the *y*-axis have a positive *x*-coordinate and

9

LINEAR FUNCTIONS AND GRAPHS

points to the left of the y-axis have a negative x-coordinate. Similarly, points above the x-axis have a positive y-coordinate and points below the x-axis have a negative y-coordinate. The points (2, –3) and (–1, 4) are shown in Figure 9.2.

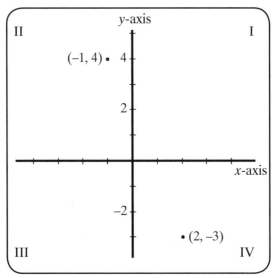

Figure 9.2.

The coordinate axes divide the plane into four parts, called **quadrants**. Quadrant I consists of those points that have positive values for both their x-coordinate and y-coordinate. Quadrant I is located in the upper right part of the plane. We continue on to Quadrants II, III, and IV in a counter-clockwise progression. Quadrant II consists of those points that have a negative value for their x-coordinate and a positive value for their y-coordinate. Quadrant III consists of those points that have negative values for both their x-coordinate and their y-coordinate. Finally, Quadrant IV consists of those points that have a positive value for their x-coordinate and a negative value for their y-coordinate. The signs for the x-coordinates and y-coordinates are summarized in this table.

Quadrant	Coordinate Signs
I	(+,+)
II	(–,+)
III	(–,–)
IV	(+,–)

Lesson 9-3 Review

1. Graph points with coordinates (–2, 4) and (2, –4).

2. Read off the coordinates of the points P and Q shown in Figure 9.3.

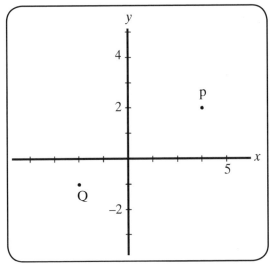

Figure 9.3.

Lesson 9-4: Slope and Intercepts

The best way to understand a function is to generate a few ordered pairs and plot them. Whenever you are trying to examine a function, it is best to create the ordered pairs in a systematic way. I recommend evaluating the function for a combination of evenly spaced negative and positive values for x. Evaluating the function when $x = 0$ is also helpful. I have evaluated the function $f(x) = 2x + 3$ for seven values of x, and have written the corresponding function values and ordered pairs in the table shown on page 178.

Notice that all of the values of x are evenly spaced and center around $x = 0$. Recall that x is called the independent variable. By "independent" we mean that we have control over what values we use

x	$f(x) = 2x + 3$	Ordered Pairs
−3	−3	(−3, −3)
−2	−1	(−2, −1)
−1	1	(−1, 1)
0	3	(0, 3)
1	5	(1, 5)
2	7	(2, 7)
3	9	(3, 9)

(within reason, of course). The function $f(x)$, or the variable y, is called the dependent variable. By "dependent" we mean that the value of the function depends on the value of x.

Once you have determined some of the ordered pairs that represent the function, I recommend graphing them. A carefully drawn picture or a graph can give lots of insight into the nature of the function. The graph of these ordered pairs is shown in Figure 9.4.

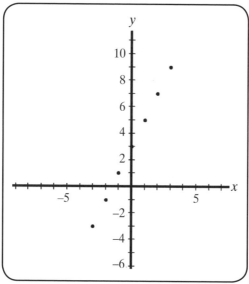

Figure 9.4.

Now that we have a picture of this function, we can make some observations. First of all, notice that all of the points appear to lie on the same line. This function was called a "linear function" and now that makes sense. I mentioned earlier that linear functions have graphs that look like lines. Now you can see that for yourself.

The graph of $f(x) = 2x + 3$ is a line that is not horizontal (as the x-axis is) and is not vertical (as the y-axis is). It is worth measuring the slant, or slope, of lines that are not vertical or horizontal. The **slope** of a line is defined as the ratio of the "rise" of a graph divided by the "run" of the graph, or the "rise over the run" of the graph. The *rise* of the graph is a measurement of the change in the dependent variable (the variable y or $f(x)$), and the *run* of the graph is a measurement of the change in the independent variable (the variable x):

$$\text{slope} = \frac{\text{change in } y}{\text{change in } x}$$

You are already comfortable using the symbol Δ to represent change, which you remember from Chapter 7. The change in a variable is just the final value of the variable minus the initial value of the variable. You usually calculate the slope of a line between two points. If you have two points, say (a, b) and (c, d), then the change in the dependent variable would be found by taking the difference between the y-coordinates of the two points: $\Delta y = d - b$. The change in the independent variable would be found by taking the difference between the x-coordinates of the two points: $\Delta x = c - a$. Then the slope of the line would be found by using the previous equation:

$$\text{slope} = \frac{\text{change in } y}{\text{change in } x} = \frac{\Delta y}{\Delta x} = \frac{d - b}{c - a}$$

You may be wondering how I knew that the point (c, d) was the final point and (a, b) was the initial point. In reality, it doesn't matter.

If I had switched them so that (a, b) was the final point and (c, d) was the initial point, I would have calculated the slope to be:

$$\text{slope} = \frac{\text{change in } y}{\text{change in } x} = \frac{\Delta y}{\Delta x} = \frac{b-d}{a-c} = \frac{(-1)(d-b)}{(-1)(c-a)} = \frac{d-b}{c-a}$$

So it is not important which point is considered to be the final point and which point is considered to be the initial point. What *is* important is that the change in *y* goes in the numerator of the fraction and the change in *x* goes in the denominator of the fraction. The other thing that is important is that you are *consistent* in which point you use as the final point and which point you use as the initial point when calculating the changes. If (a, b) is the final point when calculating Δy, it also must be the final point when calculating Δx. Let's use a couple of different points in the table to calculate the slope of the line $f(x) = 2x + 3$:

Point Number	x	$f(x) = 2x + 3$
1	−3	−3
2	−2	−1
3	−1	1
4	0	3
5	1	5
6	2	7
7	3	9

Example 1

Calculate the slope of $f(x) = 2x + 3$ using points #1 and #3.

Solution: Point #1 is the point (−3, −3) and point #3 is the point (−1, 1). The slope is:

$$\text{slope} = = \frac{\Delta y}{\Delta x} = \frac{1-(-3)}{-1-(-3)} = \frac{1+3}{-1+3} = \frac{4}{2} = 2$$

Example 2

Calculate the slope of $f(x)=2x+3$ using points #4 and #7.

Solution: Point #4 is the point (0, 3) and point #7 is the point (3, 9). The slope is:

$$\text{slope}==\frac{\Delta y}{\Delta x}=\frac{9-3}{3-0}=\frac{6}{3}=2$$

We calculated the same slope in Example 1 and in Example 2. In fact, one of the things that make lines so special is that their slope is constant. It doesn't matter which set of points you use to calculate the slope of a line.

Notice that the slope of the linear function $f(x)=2x+3$ is 2; 2 also happens to be the coefficient in front of x. Could this be a coincidence? No! You can easily recognize the slope of *any* linear function. For example, the slope of the linear function $f(x)=-3x-9$ is –3!

Example 3

Pick two points and calculate the slope of the function: $f(x)=-2x+1$

Solution: We can see from the formula for $f(x)$ that the slope is –2. It doesn't matter which two points I pick, so I will pick points corresponding to $x = 0$ and $x = 1$. The point corresponding to $x = 0$ is (0, 1) [because $f(0)=1$], and the point corresponding to $x = 1$ is (1, –1) [because $f(1)=(-2)\cdot 1+1=-1$]. The slope of the line passing through (0, 1) and (1, –1) is:

$$\text{slope}==\frac{\Delta y}{\Delta x}=\frac{-1-1}{1-0}=\frac{-2}{1}=-2$$

Example 4

Find the slope of the line that passes through the points (4, –2) and (6, 3).

Solution: Use the equation for the slope:

$$\text{slope}==\frac{\Delta y}{\Delta x}=\frac{-2-3}{4-6}=\frac{-5}{-2}=\frac{5}{2}$$

9

LINEAR FUNCTIONS AND GRAPHS

I mentioned at the beginning of the chapter that there are two important constants involved in a linear function $f(x) = mx + b$. The constant m is the slope, which we have just talked about. The other constant, b, is called the y-intercept.

The y-intercept of a function is the point where the line crosses the y-axis. At that point the value of x is 0. This point is determined by finding the value of y when you plug x into the function. If $y = mx + b$ and you substitute $x = 0$ into the equation, you will get $y = m \cdot 0 + b = b$. That is why b is called the y-intercept. The y-intercept of the line $y = mx + b$ is the point $(0, b)$.

Example 5

Find the slope and the y-intercept of the function: $f(x) = 3x + 5$

Solution: The slope is 3 and the y-intercept is the ordered pair (0, 5).

Example 6

Find the slope and the y-intercept of the function: $f(x) = 2x - 3$

Solution: The slope is 2 and the y-intercept is the ordered pair (0,–3).

Lines also have an x-intercept. The **x-intercept** of a function is the point where the line crosses the x-axis. At that point, the value of y is 0. Finding the x-intercept of a function involves setting the function equal to 0 and solving the resulting linear equation for x.

Example 7

Find the x-intercept of the function: $y = 2x - 3$

Solution: Set $y = 0$ and solve for x:

$$y = 2x - 3$$

Set $y = 0$. 　　　　　　　　　　　$0 = 2x - 3$

Add 3 to both sides. 　　　　　$0 + 3 = 2x - 3 + 3$

Simplify. 　　　　　　　　　　　$3 = 2x$

Interchange the two sides of the equation. $2x = 3$

Divide both sides by 2. $\dfrac{2x}{2} = \dfrac{3}{2}$

Simplify. $x = \dfrac{3}{2}$

The x-intercept is: $\left(\dfrac{3}{2}, 0\right)$

Lesson 9-4 Review

1. Fill in the table and then graph the points for the function:
   $f(x) = -3x + 4$

x	$f(x) = -3x + 4$
–2	
–1	
0	
1	
2	

2. Find the slope, y-intercept, and x-intercept of the function:
   $f(x) = -3x - 6$

3. Find the slope of the line that passes through the points (–2, 1) and (4, –4).

Lesson 9-5: Graphing Linear Equations

Now that we understand the idea of the slope and the intercepts of a line, it's time to get a picture of what our lines look like. The key to graphing linear functions, or lines, is that all you need is two points and a straightedge (or ruler).

In order to find two points on the line, pick two different values for x, plug them into the equation and find the corresponding values for y. Plot the two ordered pairs and connect the dots with a ruler. That's all there is to it. You can use any two points that you want.

Another way to graph a line is to find both intercepts and plot them. The way to find the x-intercept is to set $y = 0$ and solve for x;

to find the *y*-intercept set $x = 0$ and solve for *y*. I'll use the intercept method in my examples so that we can review how to find them as well as how to solve equations. The more you practice, the better you'll get. You don't have to do it my way, though. You can draw your graphs using your two favorite points and compare your graphs to mine. It doesn't matter which two points you plot and connect, as long as the points actually lie on the line in question.

Example 1

Graph the line given by the equation: $y = 2x + 1$

Solution: First, find the *y*-intercept by setting $x = 0$ and finding *y*:
$y = 2 \cdot 0 + 1 = 1$

Then find the *x*-intercept by setting $y = 0$ and solving for *x*:
$0 = 2x + 1$

$-2x = 1$

$x = -\dfrac{1}{2}$

Then plot the two intercepts $(0, 1)$ and $\left(-\dfrac{1}{2}, 0\right)$ and connect the points, as shown in Figure 9.5.

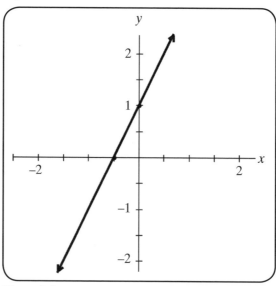

Figure 9.5.

Example 2

Graph the line given by the equation: $y = -\dfrac{3}{2}x + 2$

Solution: First, find the y-intercept by setting $x = 0$ and finding y:

$$y = -\frac{3}{2} \cdot 0 + 2 = 2$$

Then find the x-intercept by setting $y = 0$ and solving for x:

$$0 = -\frac{3}{2}x + 2$$

$$\frac{3}{2}x = 2$$

$$\frac{2}{3} \cdot \frac{3}{2}x = \frac{2}{3} \cdot 2$$

$$x = \frac{4}{3}$$

Then plot the two intercepts $(0, 2)$ and $\left(\dfrac{4}{3}, 0\right)$ and connect the points as shown in Figure 9.6.

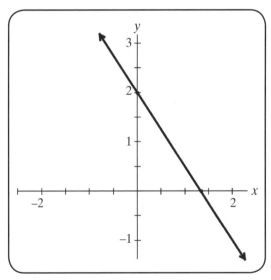

Figure 9.6.

Example 3

Graph the line given by the equation: $y = 3x$

Solution: First, find the y-intercept by setting $x = 0$: $y = 3 \cdot 0 = 0$. Notice that this line passes through the origin $(0, 0)$, which means that the x-intercept and the y-intercept are the same point. So we will need to pick a different value of x in order to come up with another point on the line. It doesn't matter what value we pick for x; to keep it simple, pick $x = 1$. Use that value to solve for y: $y = 3 \cdot 1 = 3$. Then plot the two points $(0, 0)$ and $(1, 3)$ and connect them, as shown in Figure 9.7.

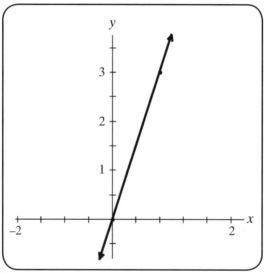

Figure 9.7.

Lesson 9-5 Review

Graph the following lines.

1. $y = -2x + 1$

2. $y = -\dfrac{1}{3}x + 3$

3. $y = \dfrac{3}{2}x$

Answer Key

Lesson 9-2 Review

1. $f(-1) = -6$
2. $f(3) = 6$
3. $f(-1) = 6$
4. $f(0) = -9$

Lesson 9-3 Review

1. The graph of the points $(-2, 4)$ and $(2, -4)$

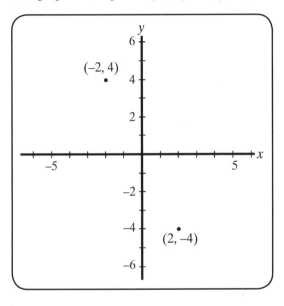

2. P is the point $(3, 2)$ and Q is the point $(-2, -1)$

Lesson 9-4 Review

1.

x	$f(x) = -3x + 4$
-2	10
-1	7
0	4
1	1
2	-2

The graph of the points in the table.

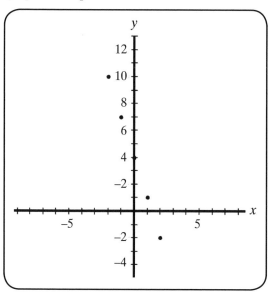

2. Slope: –3; y-intercept: $(0, -6)$; x-intercept:$(2, 0)$

3. $-\dfrac{5}{6}$

Lesson 9-5 Review

1. Graph of the line $y = -2x + 1$

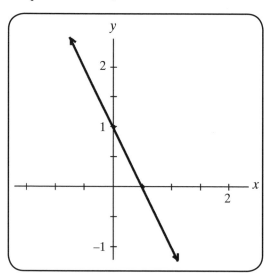

2. Graph of the line $y = -\frac{1}{3}x + 3$

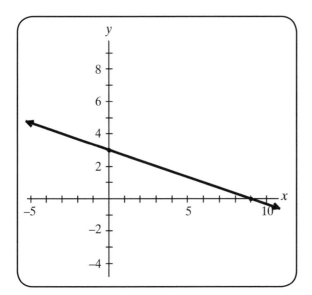

3. Graph of the line $y = \frac{3}{2}x$

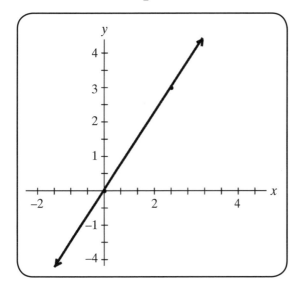

10

Algebra and Geometry

Algebra and geometry go hand-in-hand, thanks in part to Rene Descartes. Algebra provides a systematic way to approach computational problems in geometry. Before we get into the algebra of geometry, we have to discuss some basic geometric shapes. Our focus will be on the most interesting and common geometric shapes: polygons.

Lesson 10-1: Polygons

A **polygon** is a *closed* plane figure with at least three *sides*. The sides meet only at their **endpoints**. The point where two sides meet is called a **vertex**. The name of a polygon is determined by the number of sides that it has. A **triangle** is a polygon with three sides. A **quadrilateral** is a polygon with four sides. There are many different kinds of quadrilaterals. Squares, rectangles, parallelograms, and trapezoids are just a few types of quadrilaterals.

The smallest number of sides that a polygon can have is three. A three-sided polygon is called a triangle. Triangles can be classified by their *angle measures*. A triangle whose interior angles are all less than 90° is called an **acute** triangle. A triangle that has one angle that is equal to 90° is called a **right** triangle. A triangle that has one angle that is greater than 90° is called an **obtuse** triangle. A triangle whose angles all have the same measure is called an **equiangular** triangle. One important feature about triangles is that the sum of the interior angles is 180°. Figure 10.1 shows examples of acute, right, obtuse, and equiangular triangles.

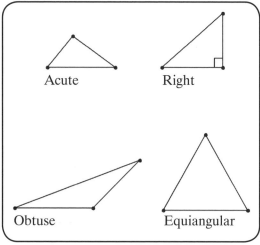

Figure 10.1.

Triangles can also be classified by the *lengths* of their sides. If all three sides of the triangle have different lengths, the triangle is **scalene**. If two sides of a triangle are the same length, the triangle is called an **isosceles** triangle. If all three sides of a triangle have the same length, the triangle is called an **equilateral** triangle. Figure 10.2 illustrates examples of scalene, isosceles, and equilateral triangles.

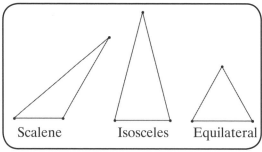

Figure 10.2.

Moving up in the realm of polygons are the four-sided polygons, or quadrilaterals. Quadrilaterals can be classified by the *relationships between their sides*. Opposite sides of a quadrilateral are parallel if, upon extending the two sides into lines, the two lines never intersect. If one pair of opposite sides of a quadrilateral are parallel, the quadrilateral is called a **trapezoid**. If *both* pairs of opposite sides of a quadrilateral are parallel, the quadrilateral is called a **parallelogram**. Every parallelogram has the property that the sides opposite each other have the same length. If, in addition to opposite sides being the same length, adjacent sides of a parallelogram have the same length, the parallelogram is called a **rhombus**. Figure 10.3 shows examples of these quadrilaterals.

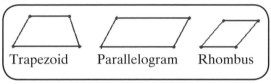

Trapezoid   Parallelogram   Rhombus

Figure 10.3.

Quadrilaterals can also be classified by the size of their *interior angles*. If all four angles of a quadrilateral are right angles, then the quadrilateral is called a **rectangle**. If all four angles of a quadrilateral are right angles *and* all four sides have the same length, the quadrilateral is a **square**. Figure 10.4 illustrates examples of these types of quadrilaterals.

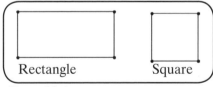

Rectangle   Square

Figure 10.4.

A rectangle, a rhombus, and a square are special types of parallelograms. Because they are parallelograms, they have the property that their opposite sides have the same length. When we describe a rectangle, we only need to specify the length of two adjacent sides.

We call these sides the **length** and **width** of the rectangle. It doesn't matter which side we call the length and which side we call the width. We can always rotate the rectangle to give it a different orientation. Fortunately, rotating a polygon (or any shape, for that matter) doesn't change the length of the sides of the polygon. It only changes its orientation and how we see it. Of course, using the word *length* to represent two ideas can be confusing. We use "length" to refer to one side of a rectangle, and we use "length" when discussing measurements in general. You should understand which usage is appropriate from the context of the problem. Don't let it bother you when I talk about the length of the length of a rectangle.

With a square, we only need to specify the length of one of its sides, as all four sides have the same length. The sides of a square are referred to as just that: sides of the square.

It is easy to introduce algebra into problems that involve geometry, as we will see in the following sections. We will start with line segments and then turn our attention to triangles and other polygons.

Lesson 10-2: Midpoint

We have already combined algebra and geometry to create the Cartesian coordinate system. This coordinate system makes graphing points and lines a breeze. This coordinate system also helps determine the midpoint of a line segment.

The **midpoint** of a line segment is the *unique* point that is exactly halfway between the two endpoints of the segment. I recommend associating the midpoint of a line segment with the *average* of the coordinates of the two points. The midpoint of the line segment with endpoints (a, b) and (c, d) is the point whose x-coordinate is the *average* of the x-coordinates of the two endpoints and whose y-coordinate is the *average* of the y-coordinates of the two endpoints. If M is the midpoint of the line segment with endpoints (a, b) and (c, d), you can use this formula to find the midpoint M:

$$M = \left(\frac{a+c}{2}, \frac{b+d}{2} \right)$$

Example 1

Find the midpoint of the line segment with endpoints (3, 8) and (–2, 6).

Solution: Using the formula for the midpoint,

$$M = \left(\frac{3+(-2)}{2}, \frac{8+6}{2} \right) = \left(\frac{1}{2}, \frac{14}{2} \right) = \left(\frac{1}{2}, 7 \right)$$

It may be helpful to see a picture of the relationship between these three points. Figure 10.5 shows these three points on the Cartesian coordinate system.

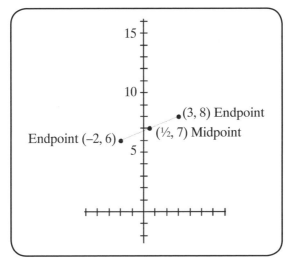

Figure 10.5.

Example 2

A line segment has an endpoint at (4, 2) and a midpoint at (1, –2). Find the location of the other endpoint.

Solution: Now we are given one endpoint and the midpoint. We need to find the coordinates of the other endpoint. Let (x, y) represent the

ALGEBRA AND
GEOMETRY

10

coordinates of the endpoint we are trying to find. From the formula for the midpoint, we see that:

$$1 = \frac{x+4}{2} \qquad\qquad -2 = \frac{y+2}{2}$$

We can solve the first equation for x and the second equation for y:

$$1 = \frac{x+4}{2} \qquad\qquad -2 = \frac{y+2}{2}$$

$$2 = x+4 \qquad\qquad -4 = y+2$$

$$x+4 = 2 \qquad\qquad y+2 = -4$$

$$x = -2 \qquad\qquad y = -6$$

So the other endpoint is located at (–2, –6).

Lesson 10-2 Review

Solve the following problems.

1. Find the midpoint of the segment with endpoints located at (–3, 5) and (2, 6).

2. If a line segment has an endpoint located at (–2, 4) and a midpoint located at (–4, 6), find the location of the other endpoint.

Lesson 10-3: Perimeter

The **perimeter** of a geometric shape is the sum of all of the lengths of the sides of the shape. For example, a triangle has three sides, and the perimeter of a triangle is the sum of the lengths of the three sides.

Example 1

Find the perimeter of the triangle shown in Figure 10.6.

Solution: The perimeter of the triangle is 5 + 3 + 6 = 14.

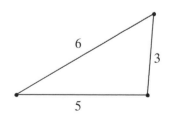

Figure 10.6.

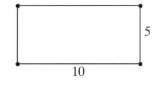
Example 2

Find the perimeter of the rectangle shown in Figure 10.7.

Solution: Opposite sides of a rectangle have the same length, so the perimeter of the rectangle is $10 + 10 + 5 + 5 = 30$.

Figure 10.7.

It's time to show you how algebra can be used to solve problems in geometry.

Example 3

The length of a rectangle is 7 inches long and the perimeter of the rectangle is 20 inches. Find the width of the rectangle.

Solution: Let x represent the width of the rectangle that we are trying to find. Because the perimeter is 20 inches, we know that $7 + 7 + x + x = 20$. We can then solve for x:

$$7 + 7 + x + x = 20$$
$$14 + 2x = 20$$
$$2x = 6$$
$$x = 3$$

So, the width of the rectangle is 3 inches long.

Example 4

The length of a rectangle is three times its width. If the perimeter of the rectangle is 64 centimeters, find the dimensions of the rectangle.

Solution: Let x represent the width of the rectangle, which is the shorter of the two sides. The length of the rectangle will be $3x$, because we were told that the length is three times its width. The perimeter of the rectangle will be $3x + 3x + x + x = 64$. We can then solve for x:

$$3x + 3x + x + x = 64$$
$$8x = 64$$
$$x = 8$$

So, the width of the rectangle is 8 centimeters long, and the length of the rectangle is 24 centimeters long. We can write the dimensions of a rectangle as either the length by the width or the width by the length. We usually write the smaller of the two first. We would write that this rectangle is 8 centimeters by 24 centimeters.

There is a very important property that applies to the perimeter of *every* triangle. It is called the triangle inequality. The **triangle inequality** states that if you add up the lengths of *any* two sides of a triangle, your answer will always be greater than the length of the third side. The triangle inequality can be used to determine whether a particular triangle can exist or not. For example, it is not possible to have a triangle with side lengths of 3, 5, and 10 because, if you add the two shorter side lengths together you will get 8, and 8 < 10. Because a triangle with side lengths of 3, 5, and 10 violates the triangle inequality, such a triangle cannot exist. You don't have to take my word for it. Try to make a triangle with these side lengths.

Lesson 10-3 Review

Solve the following problems.

1. The length of a rectangle is 15 inches and its width is 30 inches. Find the perimeter of the rectangle.

2. The length of a rectangle is 10 centimeters and the perimeter is 45 centimeters. Find the width of the rectangle.

3. The length of a rectangle is four times the width of the rectangle. The perimeter is 60 meters. Find the width of the rectangle.

4. A square has a side length of 5 inches. Find the perimeter of the square.

5. Can a triangle with sides having the lengths 10 inches, 12 inches, and 25 inches exist?

Lesson 10-4: Right Triangles and the Pythagorean Theorem

A *right* triangle is a triangle with one angle measuring 90°. The right angle can be marked with a little square to make it stand out (as in Figure 10.10 on page 202). The side opposite the right angle is called the **hypotenuse**, and the other two sides are called the **legs**. The hypotenuse will *always* be longer than either of the two sides.

The Pythagorean Theorem describes the relationship between the length of the hypotenuse and the lengths of the two legs. The **Pythagorean Theorem** states that the sum of the squares of the lengths of the legs of a right triangle will equal the square of the length of the hypotenuse. This theorem may be easier to understand if we write it algebraically. Consider the right triangle in Figure 10.8. If a and b represent the lengths of the legs of a right triangle, and if c represents the length of the hypotenuse, the Pythagorean Theorem states that:

$$a^2 + b^2 = c^2$$

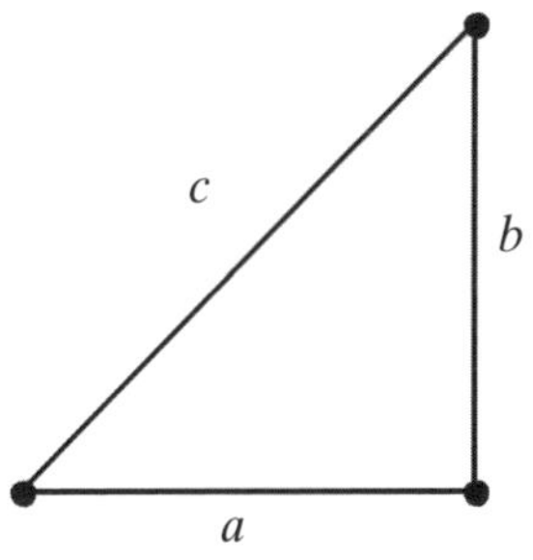

Pythagorean Theorem
$$a^2 + b^2 = c^2$$

Figure 10.8.

Many mathematicians have constructed proofs of the Pythagorean Theorem. Even non-mathematicians have discovered new ways to prove this theorem. In fact, President Garfield published a proof of the Pythagorean Theorem back in 1876.

10 ALGEBRA AND GEOMETRY

ALGEBRA AND
GEOMETRY

10

One of the nice things about the Pythagorean Theorem is that it works both ways:

- Given *any* right triangle, the lengths of the three sides satisfy the relationship $a^2 + b^2 = c^2$.

- If the lengths of the three sides of a triangle satisfy the equation $a^2 + b^2 = c^2$, where a and b are the lengths of the shorter sides and c is the length of the longest side, then the triangle is a right triangle.

The Pythagorean Theorem involves squaring lengths and solving for lengths by taking the square root of a number. Remember that if a number is a perfect square, its square root will be an integer. Otherwise, its square root will be an irrational number. We can use a calculator to approximate the square root of a number that is not a perfect square and round to get estimates for these irrational numbers.

Example 1

If the lengths of the two legs of a right triangle are 5 inches and 12 inches, find the length of the hypotenuse.

Solution: Let c represent the length of the hypotenuse and use the Pythagorean Theorem to find c.

$$5^2 + 12^2 = c^2$$
$$c^2 = 169$$
$$\sqrt{c^2} = \sqrt{169}$$
$$c = 13$$

So, the length of the hypotenuse is 13 inches.

Example 2

A right triangle has one leg that is three times longer than the other leg. If the length of the hypotenuse is 10 inches, find the lengths of the two legs.

Solution: Let a represent the length of the shorter leg. Then the length of the other leg is $3a$. Use the Pythagorean Theorem to find a:

$$a^2 + (3a)^2 = 10^2$$

$$a^2 + 9a^2 = 100$$

$$10a^2 = 100$$

$$a^2 = 10$$

$$\sqrt{a^2} = \sqrt{10}$$

$$a = \sqrt{10}$$

So the length of the shorter leg is $\sqrt{10} \approx 3.16$ inches, and the length of the longer leg is three times that, or $3\sqrt{10} \approx 9.49$. I used a calculator to approximate $\sqrt{10}$ and $3\sqrt{10}$, which are irrational numbers, and then rounded the answer.

Example 3

Sarah wants to measure the size of an angle to see if it is a right angle. She creates a triangle around it, as shown in Figure 10.9. Is Sarah's angle a right angle?

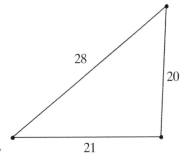

Solution: Check to see if the Pythagorean Theorem is satisfied. First, calculate the sum of the squares of the lengths of the two shorter sides:

Figure 10.9.

$$20^2 + 21^2 = 400 + 441 = 841$$

Compare that result to the square of the length of the longer side:

$$28^2 = 784$$

Because the two results are not equal, the Pythagorean Theorem does not hold, so the triangle Sarah created is not a right triangle. Therefore, Sarah's angle is not a right angle.

ALGEBRA AND GEOMETRY

10

Lesson 10-4 Review

Solve the following problems.

1. A right triangle has legs with length 4 and 6 inches. Find the perimeter of the triangle.

2. A right triangle has a hypotenuse that is twice as long as one of its legs. If the other leg is 6 inches long, find the perimeter of the triangle.

3. Is a triangle with sides of length 4, 7, and 10 centimeters a right triangle?

Lesson 10-5: Special Right Triangles

There are two right triangles that deserve special attention. The first special triangle is an isosceles right triangle. An **isosceles right triangle** is a triangle whose two legs have the same length. It can also be called a 45° - 45° - 90° triangle, because of its angle measurements. A 45° - 45° - 90° triangle is shown in Figure 10.10.

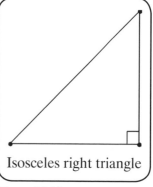

Isosceles right triangle

Example 1

Figure 10.10.

Suppose the length of the hypotenuse of an isosceles right triangle is 12 inches. Find the length of one of the legs.

Solution: With an isosceles right triangle, both legs have the same length, so it doesn't matter which leg length we find. Let *a* represent the length of one of the legs. Then the other leg also has length *a*. We can use the Pythagorean Theorem to find *a*:

$$a^2 + a^2 = 12^2$$
$$2a^2 = 144$$

$$a^2 = 72$$

$$\sqrt{a^2} = \sqrt{72}$$

$$a = \sqrt{36 \cdot 2} = \sqrt{36} \cdot \sqrt{2} = 6\sqrt{2}$$

So the length of each of the legs is $6\sqrt{2} \approx 8.49$ inches.

Example 2

The length of one leg of an isosceles right triangle is 1.5 feet. Find the length of the hypotenuse of this triangle.

Solution: The lengths of each of the legs is 1.5 feet. Let c represent the length of the hypotenuse. Use the Pythagorean Theorem to find c:

$$1.5^2 + 1.5^2 = c^2$$

$$2.25 + 2.25 = c^2$$

$$c^2 = 4.5$$

$$\sqrt{c^2} = \sqrt{4.5}$$

$$c = \sqrt{4.5} \approx 2.12$$

The length of the hypotenuse is approximately 2.12 feet.

In general, there is a special relationship between the lengths of the legs and the length of the hypotenuse in 45° - 45° - 90° triangles. If a represents the length of each of the legs and c represents the length of the hypotenuse, we see that:

$$a^2 + a^2 = c^2$$

$$2a^2 = c^2$$

$$\sqrt{2a^2} = \sqrt{c^2}$$

$$\sqrt{c^2} = \sqrt{2a^2}$$

$$c = \sqrt{a^2} \cdot \sqrt{2}$$

$$c = a\sqrt{2}$$

This relationship between the length of the hypotenuse and the length of each leg holds for *all* isosceles right triangles. This relationship is worth noticing, understanding, and becoming familiar with, but it is not necessarily worth memorizing. It is just one application of the Pythagorean Theorem.

The second special right triangle has the property that the hypotenuse is twice as long as the shorter leg. This second special right triangle doesn't have a formal name, but it is often called a 30° - 60° - 90° triangle, again because of its angle measurements. A 30° - 60° - 90° triangle is shown in Figure 10.11.

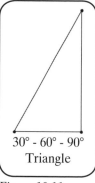

30° - 60° - 90°
Triangle

Figure 10.11.

This triangle also has a specific relationship between the lengths of the two legs. If we let a represent the length of the shorter leg, then the hypotenuse has length $2a$. This means that c, the hypotenuse, equals $2a$. Let b represent the length of the longer leg. We can use the Pythagorean Theorem to establish the relationship between the length of the longer leg and the length of the shorter leg:

$$a^2 + b^2 = c^2$$
$$a^2 + b^2 = (2a)^2$$
$$a^2 + b^2 = 4a^2$$
$$b^2 = 3a^2$$
$$\sqrt{b^2} = \sqrt{3a^2}$$
$$b = \sqrt{a^2} \cdot \sqrt{3}$$
$$b = a\sqrt{3}$$

Example 3

The length of the shorter leg of a 30° - 60° - 90° triangle is 7 centimeters. Find the length of the hypotenuse and the length of the other leg.

Solution: The length of the hypotenuse is twice the length of the shorter leg, so the length of the hypotenuse is 14 centimeters. Using the formula we just derived, the length of the longer leg is $7\sqrt{3} \approx 12.12$ centimeters.

Example 4

The length of the hypotenuse of a 30° - 60° - 90° triangle is 10 centimeters. Find the lengths of the two legs.

Solution: The length of the hypotenuse is twice the length of the shorter leg, so the length of the shorter leg is 5 centimeters. Using the formula derived earlier, the length of the longer leg is $5\sqrt{3} \approx 8.66$ centimeters.

Lesson 10-5 Review

Solve the following problems.

1. The length of one leg of an isosceles right triangle is 10 decimeters. Find the length of the hypotenuse.

2. The length of the hypotenuse of an isosceles right triangle is 16 inches. Find the length of the two legs of the triangle.

3. The hypotenuse of a 30° - 60° - 90° triangle is 24 centimeters. Find the length of the two legs.

4. The shorter leg of a 30° - 60° - 90° triangle is 30 inches. Find the lengths of the hypotenuse and the longer leg.

Lesson 10-6: The Distance Formula

If you needed to measure the distance between two points on a *number line*, all you have to do is subtract the smaller value from the larger value. For example, the distance between the points $x = 5$ and $x = 2$ is $5 - 2 = 3$. The distance between the points $x = 7$ and $x = -3$ is $7 - (-3) = 7 + 3 = 10$. The distance between the points $x = -6$ and $x = -15$ is $-6 - (-15) = -6 + 15 = 9$. Notice that the distance between two points is always a positive number.

We can use the method of calculating the distance between two points on a number line to calculate the distance between two points on *any* line. We will start by calculating the distance between two points on any *horizontal* line. If a line is horizontal, the y-coordinate of every point on the line is the same; only the x-coordinates can change. To find the distance between two points on a horizontal line, subtract the smaller x-coordinate value from the larger x-coordinate value.

Example 1

Find the distance between the points (3, 5) and (10, 5).

Solution: Notice that these two points have the same y-coordinate. These two points lie on a horizontal line, so the distance between them is the difference between the two x-coordinates: $10 - 3 = 7$.

We will now turn our attention to finding the distance between two points that lie on any *vertical* line. If a line is vertical, the x-coordinate of every point on the line is the same; only the y-coordinates can change. To find the distance between two points on a vertical line, all you have to do is subtract the smaller y-coordinate value from the larger y-coordinate value.

Example 2

Find the distance between the points (3, 5) and (3, 8).

Solution: Notice that these two points have the same x-coordinate. These two points lie on a vertical line, so the distance between them is just the difference between the two y-coordinates: $8 - 5 = 3$.

Because the distance between two points must always be a positive number, we need to make sure that we always subtract the smaller number from the larger number. If we are dealing with general points that are represented using variables, we can't tell which of the two variables represents a larger number. Fortunately, we have a way of making sure that a number is always positive: Use the absolute value!

So, if we want to make sure that an algebraic expression is always positive, we can just put absolute value symbols around it!

Now that we understand how to measure the distance between two points that lie on either a vertical or a horizontal line, we can make things more challenging by finding the distance between two points that lie on *any* kind of line. The easiest way to measure the distance between any two points is to graph them in the Cartesian coordinate system and then use the Pythagorean Theorem.

Suppose you want to measure the distance between the points $(0, 0)$ and $(3, 4)$. Graph both points on the Cartesian coordinate system. Notice that you can create a right triangle, as shown in Figure 10.12. The key is to find the coordinates of a third point that helps to create a right triangle.

The distance between our two original points is the length of the hypotenuse of this right triangle. We can find the length of each leg by calculating the vertical or horizontal distances between one of our original points and this new point. In this case, the new point is the point $(3, 0)$. Using this new point, we can find the length of each of the legs.

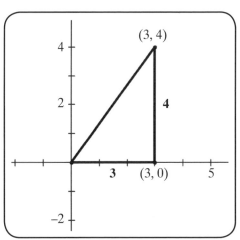

Figure 10.12.

- ⊃ The *horizontal leg* is the line segment with endpoints $(0, 0)$ and $(3, 0)$; it's length is $3 - 0 = 3$.

- ⊃ The *vertical leg* is the line segment with endpoints $(3, 4)$ and $(3, 0)$; it's length is $4 - 0 = 4$.

We have created a right triangle with two legs having lengths of 3 and 4, respectively. We can use the Pythagorean Theorem to find the length of the hypotenuse:

$$3^2 + 4^2 = c^2$$
$$9 + 16 = c^2$$
$$25 = c^2$$
$$\sqrt{c^2} = \sqrt{25}$$
$$c = 5$$

So, the length of the hypotenuse is 5. The length of the hypotenuse is the distance between the points $(0, 0)$ and $(3, 4)$, so the distance between the points $(0, 0)$ and $(3, 4)$ is 5.

You may be wondering how I knew to use the point $(3, 0)$ to help me find the distance between $(0, 0)$ and $(3, 4)$. I didn't just pull it out of thin air. I just took the first coordinate of one point and the second coordinate of the second point, and put them together to make a new point. I could have made the point $(0, 4)$, but then the right triangle that I created would have been upside down. No one likes to work on triangles that are upside down.

We can derive a general equation to find the distance between two points. We will call this equation the **distance formula**. The distance formula will enable us to calculate the distance between two points without having to draw a picture and find the (now not-so) mysterious third point. The distance formula is certainly a shortcut to solving these types of problems, and it is definitely worth remembering. But some people, myself included, find it difficult to memorize formulas. I prefer to understand where the formulas come from. That way, if I forget the formula, I can always solve the problem by using geometry and the Pythagorean Theorem.

We will derive a formula to find the distance between two points (a, b) and (c, d). First, introduce the point (c, b), as shown in Figure 10.13. This is the magical third point that will form a right triangle.

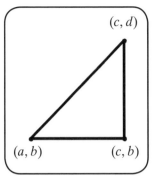

Figure 10.13.

Notice that the two points (a, b) and (c, b) lie on a *horizontal* line, because the two y-coordinates are the same. Also, the two points (c, d) and (c, b) lie on a *vertical* line, because the two x-coordinates are the same. The point (c, b) is the key to the distance formula, because it is the point that helps form the right triangle. Now we need to find the lengths of the legs of our right triangle.

Because our endpoints involve variables, there is no way that we will be able to determine which x-coordinate is bigger. We will need to use absolute value symbols to make sure that our lengths are positive.

⊃ The horizontal leg is the line segment with endpoints (a, b) and (c, b); the length of the horizontal leg is $|c - a|$.

⊃ The vertical leg is the line segment with endpoints (c, b) and (c, d); the length of the vertical leg is $|d - b|$.

Now we can use the Pythagorean Theorem to find the length of the hypotenuse, which is the distance between the points (a, b) and (c, d):

$$\left(|c - a|\right)^2 + \left(|d - b|\right)^2 = \text{distance}^2$$

$$\text{distance}^2 = (c - a)^2 + (d - b)^2$$

$$\text{distance} = \sqrt{(c - a)^2 + (d - b)^2}$$

I dropped the absolute value symbols in the second step. The reason for this is that when we square a number, whether it is positive or negative, the result will always be a non-negative number, so the absolute value symbols are no longer necessary. Keeping them around only makes the formula more complicated; because they don't add anything of value, it is not worth including them in the equation.

ALGEBRA AND
GEOMETRY
10

Looking at the distance formula, we see that the distance between two points involves squaring the difference between the x-coordinates, squaring the difference between the y-coordinates, adding the two results, and then taking the square root of that sum. There is no need to find the mysterious point, and there is no need to draw a picture. The distance formula gives the correct result *every* time.

Let me mention again the importance of understanding where the distance formula comes from: the Pythagorean Theorem. If you do get stuck and forget the distance formula, graph the two points, find the mysterious third point to create a right triangle, and use the Pythagorean Theorem.

Example 3

Find the distance between the points (2, 7) and (14, 2).

Solution: Use the distance formula. It doesn't matter which point you use first.

$$\text{distance} = \sqrt{(c-a)^2 + (d-b)^2}$$

$$\text{distance} = \sqrt{(14-2)^2 + (2-7)^2}$$

$$\text{distance} = \sqrt{(12)^2 + (-5)^2}$$

$$\text{distance} = \sqrt{144 + 25}$$

$$\text{distance} = \sqrt{169}$$

$$\text{distance} = 13$$

Lesson 10-6 Review

Find the distance between the following pairs of points.

1. (3, 5) and (1, –2)

2. (4, 1) and (2, 3)

3. (6, 3) and (8, –1)

4. (–2, 4) and (1, –2)

10 ALGEBRA AND GEOMETRY

Lesson 10-7: Area

The amount of space that a closed polygon encloses is the **area** of the polygon. Area is measured in "square units." In other words, the square is the basic unit for measuring area. Recall that a square is a quadrilateral that has four right angles and all four sides have the same length. The area of a square is defined to be the square of the length of one of its sides. A square with side length 1 inch has area 1 square inch, or 1 inch^2. A square with side length 2 centimeters has an area of 4 square centimeters, or 4 centimeters^2.

The area of a square is the motivation for calling the exponential expression a^2, "a squared." The area of a square of length a is "a **squared**," or a^2.

We use abbreviations for units of length. The most common units for length are inches, centimeters, and feet. Their abbreviations are "in," "cm," and "ft," respectively. These abbreviations for units of length will help us abbreviate units of area. A square inch is abbreviated in^2, a square centimeter is abbreviated cm^2, and a square foot is abbreviated ft^2.

Armed with a new ruler (the square unit, as in square inches) we can measure the area of other polygons besides a square. We will start with a rectangle and get more complicated. The area of a rectangle is the product of its length and its width. For example, a rectangle with length 4 inches and width 6 inches would enclose an area of 24 in^2. A rectangle with length 10 feet and width 6 feet would enclose an area of 60 ft^2.

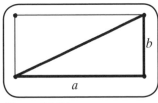

Figure 10.14.

The area of a right triangle can be related to the area of a rectangle. You can think of a right triangle as one-half of a rectangle, as shown in Figure 10.14.

Notice that the width and length of the rectangle correspond to the two legs of the right triangle. The area of the *rectangle* is then the

ALGEBRA AND GEOMETRY

10

product of the lengths of the two legs of the triangle. Because two tri-angles make up one rectangle, it follows that the area of one triangle is *one-half* of the area of the rectangle. That gives us a formula for the area of a right triangle.

> The area of a *right* triangle with leg lengths a and b is $\dfrac{a \cdot b}{2}$.

Example 1

Find the area of the polygons shown in Figure 10.15.

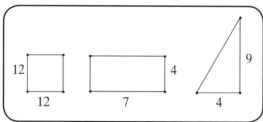

Figure 10.15.

Solution: The first polygon is a square with side length 12. The area of the square is 144 unit². The second polygon is a rectangle with length 4 and width 7. The area of the rectangle is 28 unit². The third polygon is a right triangle with legs of length 9 and 4. The area of the right triangle is:

$$\frac{9 \cdot 4}{2} = \frac{36}{2} = 18 \text{ unit}^2$$

In the previous example, I was able to find the area of a *right* triangle. Unfortunately, not every triangle is right. Before I can discuss the area of a run-of-the-mill triangle, I must define some terms. The **base** of a triangle is just one of the sides of the triangle singled out. Any side of a triangle can serve as the base. Once you specify the base of a triangle, there will be one *vertex* of the triangle that does not come in contact with the base. The distance from that vertex to the base is called the **corresponding height** of the triangle. There is no *official* height of a triangle. The height of a triangle depends on the side that serves as the base.

Now that we know about the base and the corresponding height of a triangle, we can find the area of a triangle. The area of a triangle is one-half the product of the length of the base and the corresponding height of the triangle. If b represents the length of the base and h represents the corresponding height of the triangle, the area enclosed by the triangle is given by this formula:

$$\text{Area}_{\text{triangle}} = \frac{1}{2}bh$$

Example 2

Find the areas enclosed by the triangles shown in Figure 10.16.

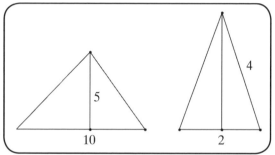

Figure 10.16.

Solution: In the first triangle, the base has length 10 and the corresponding height is 5. The enclosed area is:

$$\text{Area}_{\text{triangle}} = \frac{1}{2} \cdot 10 \cdot 5$$

$$\text{Area}_{\text{triangle}} = \frac{1}{2} \cdot 50$$

$$\text{Area}_{\text{triangle}} = 25 \text{ unit}^2$$

The base of the second triangle has length 2 and the corresponding height is 4, so the enclosed area is:

$$\text{Area}_{\text{triangle}} = \frac{1}{2} \cdot 2 \cdot 4$$

$$\text{Area}_{\text{triangle}} = 4 \text{ unit}^2$$

Lesson 10-7 Review

Find the enclosed areas of the following polygons.

1. A rectangle with width 20 decimeters and length 2 decimeters.

2. A square with length 10 centimeters.

3. An isosceles right triangle with hypotenuse of length 8 inches.

4. A triangle with base 8 inches and corresponding height 4 inches.

5. A right triangle with one leg of length 3 feet and the hypotenuse of length 5 feet.

Lesson 10-8: Circles

A **circle** is defined to be the set of all points that are an equal distance from a special point called the **center** of the circle. The distance between the center of the circle and a point on the circle is called the **radius** of the circle. An example of a circle is shown in Figure 10.17.

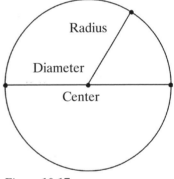

Figure 10.17.

Every line that passes through the center of a circle will intersect the circle at two distinct places. The *distance* between these two points of intersection is called the **diameter** of the circle. The diameter of a circle is related to the radius of a circle. If d represents the diameter of a circle, and r represents the radius of the circle, the diameter and the radius satisfy this equation:

$$d = 2r$$

In other words, the diameter of a circle is twice the size of the radius.

The length of the boundary of a circle is called the **circumference** of the circle. The circumference of a circle is similar to the perimeter of a polygon, in that it measures the length of the boundary.

It was observed thousands of years ago that the ratio of the circumference of a circle to its diameter is *always* the same numerical value. This really is magical. It doesn't matter whether the circle is really big, or really small—the ratio of the circumference of any circle to its diameter is always the same number. This ratio is called **pi** and is represented by the Greek letter π. Given any circle:

$$\pi = \frac{\text{circumference}}{\text{diameter}}$$

Pi is an irrational number, and its value has been calculated to more than one *trillion* decimal places. Pi appears throughout mathematics. Ancient Greek mathematicians were able to calculate π very accurately (though not to more than a trillion decimal places). This is impressive, given the technological resources available at the time. The value of π is roughly 3.141592653589; some people approximate π by the ratio $\frac{22}{7}$, but this rational approximation is just that: an approximation. Pi is *not* a rational number. Because π is a constant, we can use it to relate the circumference of a circle to its diameter, and hence its radius. If we let C represent the circumference of a circle, d represent the diameter, and r represent the radius, we have:

$$C = \pi d = 2\pi r$$

The results of calculations of the circumference of a circle are usually written as multiples of π, but I'll also give numerical approximations for the answers. Numerical approximations are usually more meaningful.

Example 1

Find the diameter and the circumference of a circle with radius 7 inches.

Solution: The diameter of a circle is twice the radius; the diameter of our circle is 14 inches. The circumference of a circle is $2\pi r$; the circumference of our circle is 14π inches. This is roughly 43.98 inches.

The area of a circle is also related to its radius. The area of a circle can be calculated using this formula:

$$\text{Area} = \pi r^2$$

The results of calculations of the area of a circle are usually written as multiples of π, but I'll also give numerical approximations for the answers. Numerical approximations are usually more meaningful.

Example 2

Find the area of a circle with radius 4 centimeters.

Solution: Use the equation for the area of a circle:
$\text{Area} = \pi (4)^2 = 16\pi \, \text{cm}^2$. This is roughly 50.27 cm².

Example 3

Find the area of a circle with a diameter of 10 feet.

Solution: In order to use our equation for area, we must know the radius of the circle. We were given the diameter of the circle, so we must use that information to find the radius. Then we can use our equation for the area of a circle:

$d = 2r$

$10 = 2r$

$r = 5$

$\text{Area} = \pi (5)^2 = 25\pi \, \text{ft}^2$

The area is roughly 78.54 ft².

Lesson 10-8 Review

Find the area and circumference of circles with the following.

1. Radius is 5 inches.

2. Radius is 16 inches.

3. Diameter is 8 inches.

4. Diameter is 10 inches.

Answer Key

Lesson 10-2 Review

1 $\left(-\frac{1}{2}, \frac{11}{2}\right)$

2. $(-6, 8)$

Lesson 10-3 Review

1. 90 inches

2. 12.5 centimeters

3. 6 meters

4. 20 inches

5. No, it violates the triangle inequality, because $10 + 12 < 25$.

Lesson 10-4 Review

1. $10 + \sqrt{52} = 10 + 2\sqrt{13}$

2. Let x represent the length one of the legs. Then the length of the hypotenuse is $2x$. Use the Pythagorean Theorem to solve for x: $x^2 + 6^2 = (2x)^2$. $x = \sqrt{12} = 2\sqrt{3}$. The sides of the triangle are 6, $2\sqrt{3}$ and $4\sqrt{3}$, so the perimeter is $6 + 6\sqrt{3}$.

3. No, because the Pythagorean Theorem is not satisfied.

Lesson 10-5 Review

1. $10\sqrt{2}$ decimeters

2. $8\sqrt{2}$ inches

3. Shorter leg: 12 centimeters; longer leg: $12\sqrt{3}$ centimeters

4. Hypotenuse: 60 inches; longer leg: $30\sqrt{3}$ inches

Lesson 10-6 Review

1. $\sqrt{53}$

2. $4\sqrt{2}$

3. $2\sqrt{5}$

4. $3\sqrt{5}$

ALGEBRA AND GEOMETRY

10

Lesson 10-7 Review

1. 40 decimeters^2

2. 100 cm^2

3. 16 in^2

4. 16 in^2

5. 6 ft^2

Lesson 10-8 Review

1. Circumference: 10π inches; area: 25π in^2

2. Circumference: 32π inches; area: 256π in^2

3. Circumference: 8π inches; area: 16π in^2

4. Circumference: 10π inches; area: 25π in^2

11

Monomials and Polynomials

Your experiences with functions have exposed you to some mathematical expressions that are very important in algebra. For example, the function $f(x) = 4x$ is the function that takes whatever is in parentheses and multiplies it by 4. The expression $4x$ is an example of a monomial. A **monomial** is an expression that involves only numbers and variables that have powers that are positive integers. A monomial in one variable is an expression of the form $k \cdot x^n$ where k is a real number and n is a positive integer. The **degree** of a monomial with only one variable is the power that the variable is raised to. A *constant* can be viewed as a monomial of the form $k \cdot x^0$ (remember from our rules for exponents that any non-zero number raised to the power of 0 is 1), so a constant is a monomial of degree 0.

A monomial with two variables is an expression of the form $k \cdot x^n \cdot y^m$. The number k is called the *coefficient* of the monomial. The **degree** of a monomial with more than one variable is the sum of the powers of all of the variables that appear. A monomial of the form $k \cdot x^n \cdot y^m$ has degree $n + m$.

Lesson 11-1: Adding and Subtracting Monomials

The only way that you can effectively add or subtract two monomials is if both of the monomials involved have the exact same variables raised to the exact same powers. The only possible variation in the monomials is in their coefficients. For example, you can add the monomials $2x^2y^3$ and $5x^2y^3$ together to get $7x^2y^3$, but if you try to add the monomials $2x^2y^3$ and $5x^2y^2$ together you will get $2x^2y^3 + 5x^2y^2$. There is no further simplification that you can do with the expression $2x^2y^3 + 5x^2y^2$ because the powers of y are different in the two monomials: The power of y in the first monomial is 3 and the power of y in the second monomial is 2. You cannot break apart a monomial. The variables in a monomial have to stick together.

Example 1

Combine the following monomials, if possible:

a. $5xy + 4xy$

b. $3x^2 - x^2$

c. $12xyz + 16xyz$

d. $12xy^2 - 5xy$

Solution:

a. $5xy + 4xy = 9xy$

b. $3x^2 - x^2 = 2x^2$

c. $12xyz + 16xyz = 28xyz$

d. $12xy^2 - 5xy = 12xy^2 - 5xy$; No simplification is possible, because the powers of y are different in the two monomials.

Lesson 11-1 Review

Combine the following monomials.

1. $4x - 6x$

2. $13xy^2 + 2xy$

3. $3xz + 5xz$

4. $6xy - 9xy$

Lesson 11-2: Multiplying and Dividing Monomials

Whenever you multiply monomials you must make use of the rules for exponents discussed in Chapter 4. Some of those rules are worth mentioning again:

$$a^m \times a^n = a^{m+n}$$

$$\frac{a^m}{a^n} = a^{m-n}$$

$$\left(a^m\right)^n = a^{m\times n}$$

Multiplying and dividing monomials is a bit more relaxed than adding and subtracting them. The variables that are involved in the monomials can be raised to different powers. All that matters is that the only operations involved in the expression are multiplication or division.

When you multiply monomials, you must multiply the constants together and then focus on each variable involved in the product. Use your rules for how to handle exponents and you can't go wrong.

Example 1

Simplify the following expressions:

a. $\left(x^3\right)\left(3x^4\right)$

b. $\left(4x^4\right)\left(3x^7\right)$

c. $\left(x^2 y^5\right)\left(-3xy^2\right)$

d. $\left(-4x^5\right)\left(-6x^2 y\right)\left(3y^2\right)$

Solution:

a. $(x^3)(3x^4) = (1 \cdot 3)(x^3 x^4) = 3x^{3+4} = 3x^7$

b. $(4x^4)(3x^7) = (4 \cdot 3)(x^4 x^7) = 12x^{4+7} = 12x^{11}$

c. $(x^2 y^5)(-3xy^2) = (1 \cdot (-3))(x^2 x)(y^5 y^2) = -3x^{2+1} y^{5+2} = -3x^3 y^7$

d. $(-4x^5)(-6x^2 y)(3y^2) = ((-4)(-6)(3))(x^5 x^2)(y^1 y^2)$

$$= 72x^{5+2} y^{1+2} = 72x^7 y^3$$

Division of monomials proceeds in a similar manner. Check for common factors of the constant terms and cancel wherever possible. Then focus on the variables involved in the quotient. Use your rules for dividing exponential expressions to simplify.

Example 2

Simplify the following expressions:

a. $\dfrac{2x^4}{x^2}$

b. $\dfrac{36x^6}{9x^4}$

Solution:

a. $\dfrac{2x^4}{x^2} = \left(\dfrac{2}{1}\right)\left(\dfrac{x^4}{x^2}\right) = 2x^{4-2} = 2x^2$

b. $\dfrac{36x^6}{9x^4} = \left(\dfrac{36}{9}\right)\left(\dfrac{x^6}{x^4}\right) = 4x^{6-4} = 4x^2$

When you have a combination of multiplication and division in one expression, simplify the numerator and denominator separately and then do the division.

Example 3

Simplify the following expressions. Write your answer in terms of positive exponents.

a. $\dfrac{\left(x^3 y^4\right)\left(2x^2 y^3\right)}{-xy^2}$

b. $\dfrac{\left(-4x^4\right)\left(-6xy\right)}{3y^2}$

c. $\dfrac{\left(6x^8 y^5\right)\left(2x^{-2} y^3\right)}{\left(-x^3 y^2\right)\left(3xy\right)}$

Solution:

a. $\dfrac{\left(x^3 y^4\right)\left(2x^2 y^3\right)}{-xy^2} = \dfrac{(1 \cdot 2)\left(x^3 x^2\right)\left(y^4 y^3\right)}{-xy^2} = \dfrac{2x^{3+2} y^{4+3}}{-xy^2} = \dfrac{2x^5 y^7}{-xy^2}$

$$= -2x^{5-1} y^{7-2} = -2x^4 y^5$$

b. $\dfrac{\left(-4x^4\right)\left(-6xy\right)}{3y^2} = \dfrac{(-4)(-6)\left(x^4 x\right)y}{3y^2} = \dfrac{24x^{4+1} y}{3y^2} = \dfrac{24x^5 y}{3y^2}$

$$= 8x^5 y^{1-2} = 8x^5 y^{-1} = \dfrac{8x^5}{y}$$

c. $\dfrac{\left(6x^8 y^5\right)\left(2x^{-2} y^3\right)}{\left(-x^3 y^2\right)\left(3xy\right)} = \dfrac{(6 \cdot 2)\left(x^8 x^{-2}\right)\left(y^5 y^3\right)}{((-1)(3))\left(x^3 x\right)\left(y^2 y\right)} = \dfrac{12x^6 y^8}{-3x^4 y^3} = -4x^2 y^5$

Lesson 11-2 Review

Simplify the following expressions. Your answer should only contain positive exponents.

1. $\left(2x^3\right)\left(4x^8\right)$

2. $\left(-4x^3 y^4\right)\left(-3x^4 y^7\right)$

3. $\dfrac{-9x^6}{12x^3}$

4. $\dfrac{50x^{13} y^7}{15x^8 y^3}$

5. $\dfrac{\left(4xy^4\right)\left(10x^3 y^2\right)}{3x^3 y^2}$

6. $\dfrac{\left(6x^2y^5\right)\left(-4x^5y^3\right)}{\left(-10xy^2\right)\left(3x^3y\right)}$

Lesson 11-3: Adding and Subtracting Polynomials

A polynomial is the sum of two or more monomials. The sign of the constant k involved in each of the monomials determines whether the monomial is added to, or subtracted from, the other monomials in the chain. The monomials that make up a polynomial are usually written in descending order based on the degree of the monomial. In other words, you start writing a polynomial by writing the monomial with the highest degree first, then the second highest degree, and so on. This is considered the **standard form** of a polynomial. When a polynomial is written in standard form, the coefficient of the first term is called the **leading coefficient**.

The **degree** of a polynomial is simply the largest of the degrees of the monomials that make up the polynomial.

Some polynomials have special names. A polynomial made up of two monomials is called a **binomial**. A polynomial made up of three monomials is called a **trinomial**. The following table gives examples of some of the more common polynomials that you will encounter in algebra.

Polynomial	Leading Coefficient	Degree	Common Name	Type of Polynomial
4	4	0	Constant	Monomial
$2x$	2	1	Linear	Monomial
$3x - 9$	3	1	Linear	Binomial
x^2	1	2	Quadratic	Monomial
$x^2 + x$	1	2	Quadratic	Binomial
$x^2 + 3x - 9$	1	2	Quadratic	Trinomial
$-2x^3$	-2	3	Cubic	Monomial
$-5x^3 + 3x + 1$	-5	3	Cubic	Trinomial

To add or subtract polynomials, you will need to apply the rules for adding or subtracting monomials. Remember that you can only combine monomials that involve the same variables raised to the same powers. So, when you try to add or subtract polynomials, you have to look for monomials that match up and combine them. When you are subtracting one polynomial from another polynomial, you must distribute the negative sign *throughout* the polynomial that is being subtracted.

Example 1

Simplify the following expressions:

a. $\left(x^2 + 3x - 2\right) + \left(3x^2 - 5x - 6\right)$

b. $\left(3x - 8y\right) + \left(5x + 2y\right)$

c. $\left(x^2y - 3x\right) - \left(4x^2y + 3x\right)$

d. $\left(x^2 + 3x + 2\right) - \left(2x^2 - 5x - 6\right)$

e. $\left(3x - 8y\right) - \left(x + 2y\right)$

Solution:

a. $\left(x^2 + 3x - 2\right) + \left(3x^2 - 5x - 6\right) = 4x^2 - 2x - 8$

b. $\left(3x - 8y\right) + \left(5x + 2y\right) = 8x - 6y$

c. $\left(x^2y - 3x\right) - \left(4x^2y + 3x\right) = \left(x^2y - 3x\right) - 4x^2y - 3x = -3x^2y - 6x$

d. $\left(x^2 + 3x + 2\right) - \left(2x^2 - 5x - 6\right) = \left(x^2 + 3x + 2\right) - 2x^2 + 5x + 6$
$$= -x^2 + 8x + 8$$

e. $\left(3x - 8y\right) - \left(x + 2y\right) = \left(3x - 8y\right) - x - 2y = 2x - 10y$

Lesson 11-3 Review

Simplify the following expressions.

1. $\left(3x^2 - 2x + 1\right) + \left(x^2 + 4x + 8\right)$

2. $\left(3x^2 + 2x + 1\right) - \left(2x^2 - 4x - 5\right)$

3. $(6x+2y)+(2x-y)$

4. $(3x+7y)-(2x-y)$

Lesson 11-4: Multiplying Polynomials

In order to understand how to multiply two polynomials together, it is best to start with how to multiply a monomial and a polynomial. The key to multiplying a monomial and a polynomial together is to use the distributive property and the rules for multiplying numbers with the same base.

Example 1

Find the product: $6(2x-8)$

Solution: $6(2x-8)=6\cdot2x+(6)(-8)=12x-48$

Example 2

Find the product: $x(2x+3)$

Solution: $x(2x+3)=x(2x)+(x)(3)=2x^2+3x$

Example 3

Find the product: $3x^2(2x^3-8)$

Solution: $3x^2(2x^3-8)=(3x^2)(2x^3)+(3x^2)(-8)=6x^5-24x^2$

When you multiply two binomials, you need to use the distributive property twice, making sure that you find all of the necessary products:

$$(a+b)(c+d)=a(c+d)+b(c+d)=ac+ad+bc+bd$$

It doesn't matter whether the binomials involve addition or subtraction. You distribute the sign:

$$(a-b)(c+d)=a(c+d)-b(c+d)=ac+ad-bc-bd$$

$$(a+b)(c-d)=a(c-d)+b(c-d)=ac-ad+bc-bd$$

$$(a-b)(c-d) = a(c-d) - b(c-d) = ac - ad - bc + bd$$

Example 4

Find the product: $(x+3)(x+2)$

Solution: Use the distributive property twice and gather like terms:

$$(x+3)(x+2) = x(x+2) + 3(x+2)$$
$$= x^2 + 2x + 3x + 6$$
$$= x^2 + 5x + 6$$

Example 5

Find the product: $(x+3)(2x-4)$

Solution: Use the distributive property twice and gather like terms:

$$(x+3)(2x-4) = x(2x-4) + 3(2x-4)$$
$$= 2x^2 - 4x + 6x - 12$$
$$= 2x^2 + 2x - 12$$

Example 6

Find the product: $(2x-3)(x+1)$

Solution: Use the distributive property twice and gather like terms:

$$(2x-3)(x+1) = 2x(x+1) - 3(x+1)$$
$$= 2x^2 + 2x - 3x - 3$$
$$= 2x^2 - x - 3$$

Example 7

Find the product: $(x-3)(x-5)$

Solution: Use the distributive property twice and gather like terms:

$$(x-3)(x-5) = x(x-5) - 3(x-5)$$
$$= x^2 - 5x - 3x + 15$$
$$= x^2 - 8x + 15$$

11

MONOMIALS AND POLYNOMIALS

The process of multiplying binomials together is often referred to as **FOIL**, which stands for first, outside, inside, and last. This represents one way of doing the multiplication as shown in Figure 11.1.

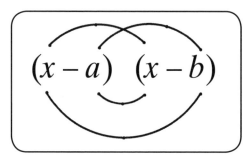

Figure 11.1.

Lesson 11-4 Review

Find the following products.

1. $3x^3\left(x^2+2x+1\right)$
2. $(x+4)(x-1)$
3. $(x-3)(x+2)$
4. $(2x-1)(x+3)$
5. $(x-2)(x+2)$
6. $(x-1)\left(x^2+x+1\right)$

Answer Key

Lesson 11-1 Review

1. $-2x$

2. $13xy^2 + 2xy$; This cannot be combined because the powers of y do not match.

3. $8xz$

4. $-3xy$

Lesson 11-2 Review

1. $8x^{11}$

2. $12x^7y^{11}$

3. $-\dfrac{3x^3}{4}$

4. $\dfrac{10x^5y^4}{3}$

5. $\dfrac{40xy^4}{3}$

6. $\dfrac{4x^3y^5}{5}$

Lesson 11-3 Review

1. $4x^2 + 2x + 9$

2. $x^2 + 6x + 6$

3. $8x + y$

4. $x + 8y$

Lesson 11-4 Review

1. $3x^5 + 6x^4 + 3x^3$

2. $x^2 + 3x - 4$

3. $x^2 - x - 6$

4. $2x^2 + 5x - 3$

5. $x^2 - 4$

6. $x^3 - 1$

12

Applications

Algebra has been around for a long time pre-cisely because it is so useful. Algebra provides us with efficient techniques that enable us to solve problems systematically. Up until now, I have given you the equations to solve. At this point in the book, the problems will be put into the context of a story and *you* will have to come up with an equation first, and then solve it.

Lesson 12-1: How to Approach Word Problems

A word problem involves a description of a situation and a question to answer. There are no variables in the word problem; it is your job to come up with variables and write an equation based on the information given in the original question. The nice thing about word problems is that they are a closer representation of how algebra will serve you in everyday life. I seriously doubt that you will be walking down the street and someone will ask you, "What is $x \div 5$ if $x = 20$?" However, you may encounter someone who will ask you if you if you can change a twenty-dollar bill into five-dollar bills, and you would need to figure out how many five-dollar bills to give him or her for the twenty.

I recommend using a systematic method for approaching and solving word problems. Here is one problem-solving approach that will help you solve word problems.

➲ **Read the problem description and pick out the important information.** Determine what you are given and what you are asked to find. Choose variables to represent what you are given and what you need to find. You are creating the equations, so you can choose the variables. Don't limit yourself to the variables x and y; sometimes it's helpful to use the first letter of what the variable represents. If you get too deep into a problem it is easy to forget what your variables mean. It may be worth making a note of it somewhere.

➲ **Read the word problem and interpret the description of the problem in terms of your variables.** Be sure to pay attention to the units involved. There are many ways to represent addition, subtraction, multiplication, and division. Words such as *sum, more, total, and, plus,* or *increase* indicate that terms should be added. Words such as *difference, less than, fewer, between,* or *decreased* indicate that terms should be subtracted. Multiplication is often referred to using words including *product, times, twice, percent,* or *of.* Division will be involved when you see such words as *ratio, divided by, half, third,* or *quotient.* The word *is* and the phrase *the same as* represent equality.

➲ **Solve the equation or evaluate the expression for a particular value of one of your variables.** Use the problem-solving strategies discussed in the earlier chapters to help you solve equations. Be sure to check your answers to catch any mistakes.

⊃ **Re-read the question and be sure that the answer you found actually addresses the question asked.** You should think about your answer to make sure that it makes sense. Giving an absurd answer to a question is a dead give-away that you didn't check your work.

If you use this approach to solve word problems, you should have as much success solving word problems as I have.

Lesson 12-2: Finding Integers

In this type of problem, you are given clues about an integer and you are asked to find the integer. Pay special attention to the various ways to represent addition, subtraction, multiplication, and division. Always remember the order of operations.

Example 1

Five more than an integer is the same as twice the integer. Find the integer.

Solution: Let x represent the integer. From the problem description, we can write down an equation and solve it:

$5 + x = 2x$

$5 = x$

So, the integer is 5.

Example 2

Two more than twice an integer is 8. Find the integer.

Solution: Let x represent the integer. From the problem description, we can write down an equation and solve it:

$2 + 2x = 8$

$2x = 6$

$x = 3$

The integer is 3.

12 APPLICATIONS

Example 3

Suppose you take an integer, subtract 8, multiply by 7, add 10, and divide by 5. If the result is 9, what was the original integer?

Solution: Let x represent the original integer. From the problem description we can write down an equation and solve it. When you write down the equation you will need to keep in mind the order of operations. Usc parcntheses where necessary:

$$\frac{(x-8)\cdot 7+10}{5} = 9$$

Multiply both sides of the equation by 5. $5\cdot\dfrac{(x-8)\cdot 7+10}{5} = 5\cdot 9$

Simplify. $(x-8)\cdot 7+10 = 45$

Subtract 10 from both sides. $(x-8)\cdot 7+10-10 = 45-10$

Simplify. $(x-8)\cdot 7 = 35$

Divide both sides of the equation by 7. $\dfrac{(x-8)\cdot 7}{7} = \dfrac{35}{7}$

Simplify. $x-8 = 5$

Add 8 to both sides of the equation. $x-8+8 = 5+8$

Simplify. $x = 13$

The integer is 13.

Lesson 12-2 Review

Solve the following problems.

1. Six more than 5 times an integer is 41. Find the integer.

2. Five less than twice an integer is the same as 3 times the integer. Find the integer.

3. Take an integer and add 2, multiply the result by 5 and divide that result by 3. If the result is 10, find the original integer.

Lesson 12-3: Rate Problems

Rate problems revolve around the idea that rate times time equals distance. If r represents the rate, t is the time, and d represents the distance, then this equation can be written $r \cdot t = d$. You are usually given values for two of the three quantities and are asked to find the third.

Example 1

Amy plans to drive 520 miles to Port Orange, Florida, to visit with her nephew. If she drives at a rate of 65 miles per hour, how long will it take her to make the trip?

Solution: Start with the equation $r \cdot t = d$. We know the distance (520 miles) and the rate (65 miles per hour) so we can find the time:

$$r \cdot t = d$$

Substitute in for the rate and the distance. $\quad 65 \cdot t = 520$

Divide both sides by 65.

$$t = \frac{520}{65}$$

Simplify. $\quad t = 8$

It will take her 8 hours to make the trip.

Example 2

The sound of thunder travels about 1 mile in 5 seconds. Suppose that a bolt of lightning strikes and you hear the thunder after 8 seconds. How far away did the lightning strike?

Solution: The rate that the thunder travels is $\dfrac{1 \text{ mile}}{5 \text{ seconds}}$, and the time that it took the thunder to reach your ears is 8 seconds. Use the equation $r \cdot t = d$ to solve for the distance:

$$r \cdot t = d$$

$$\frac{1}{5} \cdot 8 = d$$

$$d = \frac{8}{5} = 1.6$$

The lightning struck 1.6 miles away from you.

12 APPLICATIONS

Example 3

The 2001 *Mars Odyssey* was launched April 7, 2001, and arrived October 24, 2001. If the *Odyssey* traveled a total distance of 286 million miles, how many miles per hour was the spacecraft traveling?

Solution: In order to find the rate of the spacecraft in miles per hour, we need to know how many hours the *Odyssey* traveled. Between April 7th and October 24th, there are 191 days. There are 24 hours in a day, so the *Odyssey* traveled 4,584 hours. If *r* represents the rate of the *Odyssey*, we can start with the rate equation:

$$r \cdot t = d$$

Substitute in for the distance and the time. $\quad r \cdot 4,584 = 286,000,000$

Divide both sides of the equation by 4,584. $\quad \dfrac{4,584}{4,584}r = \dfrac{286,000,000}{4,584}$

Simplify. $\qquad\qquad\qquad\qquad\qquad\qquad r = 62,391$

The *Odyssey* traveled roughly 62,391 miles per hour.

Lesson 12-3 Review

Solve the following problems.

1. Nancy flew 1,040 miles to visit her sister in New York. If her plane is scheduled to take off at 7:00 a.m. and will arrive at 10:15 a.m., find the average speed that the plane will travel in miles per hour.

2. Nancy's sister lives 32 miles from the airport and would like to meet her sister at the gate. If Nancy's flight is scheduled to arrive at 10:15 a.m. and traffic is light enough that her sister can drive 48 miles per hour, what time should she leave for the airport?

3. Harry's car has a flat tire, so he has to ride his bicycle to work. His office is 2 miles away from his house and he has to get to work by 8:00 a.m. If he leaves his house at 7:54 a.m., how fast does he have to ride his bicycle in order to get to work on time? Give his rate in miles per hour.

Lesson 12-4: Money Problems

Problems that involve money often involve either different coins or bills. If you work in dollars (when the problems involve bills) or cents (if the problem involves coins), you will usually end up dealing with whole numbers instead of fractions or decimals.

Example 1

Amanda's cell phone plan costs $30 per month plus $0.35 per minute that she talks more than 250 minutes. If she talked for 320 minutes in September, what will her phone bill be?

Solution: Let m represent the number of minutes that Amanda talks on her phone, and let B represent the bill. We have the following equation:

$B = 30 + 0.35(m - 250)$

We know that m is 320, so we have to solve for B when $m = 320$:

$B = 30 + 0.35(320 - 250)$

$B = 30 + 0.35(70)$

$B = 30 + 24.5$

$B = 54.5$

Amanda's phone bill will be $54.50.

Example 2

Nathan just purchased a home entertainment center. His plasma TV cost 31 times as much as his DVD player cost. If his DVD player cost $90, how much did his plasma TV cost?

Solution: Let d represent the cost of the DVD player, and let P represent the cost of the plasma TV. Then we have the following equation relating the two variables:

$P = 31d$

12

APPLICATIONS

We need to evaluate P when $d = 90$:

$P = 31 \cdot 90 = 2{,}790$

Nathan paid $2,790 for his plasma TV.

Example 3

An electric company charges a customer a $10 service charge plus $0.04 for each kWh of electricity used. Find the amount Robert will have to pay if he uses 1,500 kWh in a month.

Solution: Let w represent the kWh used, and let B represent the monthly bill. We have the following equation:

$B = 10 + 0.04w$

We need to evaluate this equation when $w = 1{,}500$:

$B = 10 + 0.04w$

$B = 10 + 0.04 \cdot 1{,}500$

$B = 10 + 60$

$B = 70$

Robert's electric bill will be $70.

Example 4

Cameron spent $\frac{2}{3}$ of his money on video games. He has $30 left. How much did he have to start with?

Solution: Let m represent the money Cameron spent on video games, and let S represent the money he started with. The money he started with is split up into two groups: the money he has left and the money he spent on games. He spent $\frac{2}{3}$ of the money he started with on games. We can write the following equation:

$$S = 30 + \frac{2}{3}S$$

We can now solve for S:

$$S = 30 + \frac{2}{3}S$$

Subtract $\frac{2}{3}S$ from both sides. $S - \frac{2}{3}S = 30 + \frac{2}{3}S - \frac{2}{3}S$

Simplify.

$$\frac{1}{3}S = 30$$

Multiply both sides by 3. $3 \cdot \frac{1}{3}S = 3 \cdot 30$

Simplify. $S = 90$

Cameron started out with $90.

Lesson 12-4 Review

Solve the following problems.

1. Erich spends $\frac{1}{5}$ of his weekly salary on food and $\frac{1}{4}$ of his weekly salary on rent. If, after paying his rent and buying his food for the week he has $110 of his weekly paycheck left over, how much does Erich earn each week?

2. The *Port Orange Press* pays its sales force a base salary plus commission. The base salary is $200 per week and 10% commission on all sales in excess of $1,000. What will an employee earn if his or her sales level is $2,500 for one week?

Lesson 12-5: Spatial Problems

Spatial problems combine concepts in geometry with algebraic problem-solving skills to answer questions. Being familiar with equations to find the perimeter, circumference, and area of special shapes such as triangles, rectangles, squares, and circles will come in handy.

Example 1

One leg of an isosceles right triangle is 15 inches long. Find the perimeter of the triangle.

12 APPLICATIONS

Solution: In order to find the perimeter of the triangle we must find the lengths of all three sides. We already know the lengths of two of the three sides (the two legs), so we need to find the length of the hypotenuse. Let c represent the length of the hypotenuse. Then, as we discussed in Chapter 10, $c = 15\sqrt{2}$ inches.

The perimeter of the triangle is $15 + 15 + 15\sqrt{2} = 30 + 15\sqrt{2}$, or roughly 51.21 inches.

Example 2

The length of the hypotenuse of a 30° - 60° - 90° triangle is 6 centimeters. Find the area of the triangle.

Solution: The area of a right triangle is one-half the product of the lengths of the two legs. We know that the hypotenuse is 6 centimeters, so the length of the shorter side is 3 centimeters, and the length of the longer leg is $3\sqrt{3}$ centimeters.

$$Area_{triangle} = \frac{1}{2} \cdot 3 \cdot 3\sqrt{3} = \frac{9}{2}\sqrt{3} \, cm^2$$

This is roughly 7.79 cm².

Example 3

Scott charges $2.10 per square foot to tile a room. How much will I have to pay Scott to tile my 12-foot-by-10-foot rectangular dining room?

Solution: The area of the rectangular dining room is 12×10, or 120 ft². Let x represent the amount that Scott will charge to tile the dining room. Scott charges $2.10 per ft², so we can set up a proportion:

$$\frac{\$2.10}{1\,ft^2} = \frac{\$x}{120\,ft^2}$$

We can cross-multiply and solve for x:

$$2.10 \cdot 120 = x \cdot 1$$

$$x = 252$$

Scott will charge $252 to tile the dining room.

Example 4

Dan needs to replace the siding on the front of his house. In order to calculate how much siding he needs, he drew the schematic diagram shown in Figure 12.1. Based on this diagram, how much siding will Dan need?

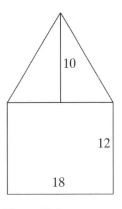

Solution: To calculate the total amount of siding needed, we need to divide the house into two sections: the rectangular section and the triangular section. The area of the rectangular section is:

Figure 12.1.

$$\text{Area}_{\text{rectangle}} = 12 \cdot 18 = 216$$

The area of the triangular section is one-half of the product of the length of the base and the corresponding height. In this case:

$$\text{Area}_{\text{triangle}} = \frac{1}{2} bh = \frac{1}{2} \cdot 18 \cdot 10 = 90$$

The total area is the area of each section:

$$\text{Area}_{\text{total}} = \text{Area}_{\text{rectangle}} + \text{Area}_{\text{triangle}} = 216 + 90 = 306$$

The total area is 306 ft².

Lesson 12-5 Review

Solve the following problems.

1. The two legs of a right triangle are 5 inches and 12 inches. Find the perimeter of the triangle.

2. The length of the shorter leg of a 30° - 60° - 90° triangle is 6 centimeters. Find the area of this triangle.

3. Find the area of a circle inscribed in a square of length 10 decimeters, as shown in Figure 12.2.

Figure 12.2.

12

APPLICATIONS

Answer Key

Lesson 12-2 Review

1. 7; the equation is $5x + 6 = 41$

2. −5; the equation is $2x - 5 = 3x$

3. 4; the equation is $\dfrac{5(x+2)}{3} = 10$

Lesson 12-3 Review

1. 320 miles per hour; the distance is 1,040 miles, and the time is 3.25 hours.

2. 9:35; the rate is 48 miles per hour, and the distance is 32 miles.

3. 20 miles per hour; the distance is 2 miles, and the time is 0.1 hours.

Lesson 12-4 Review

1. Let m represent Erich's salary. The equation that models this situation is $m - \frac{1}{5}m - \frac{1}{4}m = 110$. Solve this equation for m: Erich earns $200 per week.

2. Let m represent the weekly salary, and let s represent the weekly sales. The equation that models this situation is $m - \frac{1}{5}m - \frac{1}{4}m = 110$. Evaluate the expression on the right when $s = 2,500$: The weekly salary is $350.

Lesson 12-5 Review

1. Use the Pythagorean Theorem to find the length of the hypotenuse. The length of the hypotenuse is 13 inches. The perimeter of the triangle is 30 inches.

2. The length of the longer leg is $6\sqrt{3}$ centimeters, so the area of the triangle is $18\sqrt{3}$.

3. The radius of the circle is 10 decimeters, so the area of the circle is 25π decimeters².

Final Exam

1. Which of the following statements is FALSE?
   a. 432 is evenly divisible by 9
   b. 12 and 35 are relatively prime
   c. The remainder of $51 \div 5$ is $\frac{1}{5}$
   d. The least common multiple of 12 and 20 is 60
   e. Both addition and multiplication commute

2. The greatest common divisor of 308 and 165 is the same as which of the following?
   a. The least common multiple of 42 and 35
   b. The greatest common divisor of 42 and 35
   c. The greatest common divisor of 330 and 143
   d. 4,620
   e. None of the above

3. Simplify the expression: $4\left[12 - 3(8 - 5)\right] - 1$
   a. 11
   b. 24
   c. 108
   d. 36
   e. None of the above

4. Simplify the expression: $35 \div \left(6 - |-1|\right)$
   a. 5
   b. 7

   c. 4.83

   d. 6.83

   e. None of the above

5. Simplify the expression: $\dfrac{1}{2} \times \left(\dfrac{1}{\frac{1}{2}} + \dfrac{1}{\frac{1}{3}} \right)$

   a. $\dfrac{5}{6}$

   b. $\dfrac{5}{12}$

   c. $\dfrac{5}{2}$

   d. $\dfrac{3}{5}$

   e. None of the above

6. Simplify: $\left(4u^2v^4\right)\left(3u^3v\right)$

   a. $7u^5v^5$

   b. $12u^5v^5$

   c. $12u^6v^4$

   d. $7u^6v^4$

   e. None of the above

7. Simplify $\dfrac{6y^2}{14xy} \times \dfrac{7xy}{4x^2}$

   a. $\dfrac{3y^2}{4x^3}$

   b. $\dfrac{3y^2}{4x^2}$

   c. $\dfrac{42xy^3}{56x^3y}$

   d. $\dfrac{4x^2}{3y^2}$

   e. None of the above

8. Simplify: $\left(\dfrac{5xy^2}{4z^3}\right)^2$

   a. $\dfrac{25x^2y^4}{16z^6}$

   b. $\dfrac{5x^2y^4}{4z^6}$

   c. $\dfrac{10x^2y^4}{8z^6}$

   d. $\dfrac{25x^3y^4}{16z^5}$

   e. None of the above

9. A small order of General Tso's Chicken costs $6.45, a large order costs $8.35, and delivery costs $3.00. How much will it cost to have three small orders and two large orders delivered?

   a. $37.95

   b. $36.05

   c. $40.95

   d. $39.05

   e. None of the above

10. Bob ordered 4 large pizzas for a party. Each pizza is cut into 8 pieces, and Bob invited 5 friends. If everyone has the same number of slices, how many slices of pizza will be left over?

   a. 1 slice

   b. 2 slices

   c. 3 slices

   d. 4 slices

   e. None of the above

11. Brandon wants to make three loaves of bread. Each loaf requires $2\frac{3}{4}$ cups of flour. How many cups of flour will Brandon need to make all three loaves?

   a. $7\frac{3}{4}$ cups

b. 8 cups

c. $8\frac{1}{4}$ cups

d. $8\frac{1}{2}$ cups

e. None of the above

12. Brad, Alpay, Shahin, and Christina are playing a game. The person with the most money (in change) gets to keep everyone else's money. Brad has 7 quarters, 3 dimes, 4 nickels and 6 pennies. Alpay has 6 quarters, 4 dimes, 7 nickels, and 2 pennies. Shahin has 8 quarters, 1 nickel, and 2 pennies. Christina has 6 quarters, 3 dimes, 8 nickels, and 8 pennies. Who wins the game?

a. Brad

b. Alpay

c. Shahin

d. Christina

e. None of the above

13. Solve the following equation for x: $5x + 12 = -3x + 60$

a. $x = -4$

b. $x = 6$

c. $x = -6$

d. $x = 8$

e. None of the above

14. Evaluate $(3a)^2 - 2b$ when $a = 2$ and $b = 4$

a. 4

b. 12

c. 28

d. 30

e. None of the above

15. One cubic centimeter of pure silver has a mass of 10.5 grams. If one pound corresponds to a mass of 454 grams, what would a 727.9 cubic centimeter silver bar weigh? Round your answer to the hundredths' place.

a. 7,642.95 pounds

b. 16.83 pounds

c. 6.55 pounds

d. 65.5 pounds

e. None of the above

16. What is the slope of the line that passes through the points (2, 7) and (6, 23)?

a. 4

b. –4

c. $\dfrac{1}{4}$

d. $-\dfrac{1}{4}$

e. None of the above

17. Rachael bought a sweater on sale for $34.99. The original price of the sweater was $49.99. What was percent change in the price of the sweater?

a. –30%

b. –43%

c. –70%

d. 43%

e. None of the above

18. Solve the following inequality: $5x - 7 > 2x + 19$

a. $x > \dfrac{7}{26}$

b. $x > 4$

c. $x > \dfrac{26}{3}$

d. $x > \dfrac{26}{7}$

e. None of the above

19. Solve the following inequality x: $3(2x-7) < 5(x+2)$

   a. $x < 17$

   b. $x < 9$

   c. $x < 26$

   d. $x < 31$

   e. None of the above

20. Find the coordinates of the point located three units to the right of the y-axis and two units above the x-axis.

   a. $(-3, -2)$

   b. $(3, 2)$

   c. $(-2, 3)$

   d. $(2, 3)$

   e. None of the above

21. Which of the following three ordered pairs satisfy the equation $y = x + 3$?

   a. $(-3, 3)$, $(-2, 1)$, $(-1, 2)$

   b. $(-3, 0)$, $(-2, -2)$, $(-1, 2)$

   c. $(-3, 0)$, $(-2, 1)$, $(-1, 2)$

   d. $(-2, 1)$, $(-1, 2)$, $(0, -3)$

   e. None of the above

22. Find the slope and y-intercept of the line $3y + 24x = 15$.

   a. Slope: -24, y-intercept: 15

   b. Slope: -5, y-intercept: 5

   c. Slope: 5, y-intercept: -8

   d. Slope: -8, y-intercept: 5

   e. None of the above

23. Subtract $5x^2 - 9$ from $7x^2 - x + 1$:

   a. $2x^2 - x + 10$

   b. $-2x^2 + x - 10$

   c. $12x^2 - x - 8$

d. $2x^2 - x - 8$

e. None of the above

24. Simplify: $2u(3u+4)+3u(2u+1)$

   a. $6u^2 + 5u$

   b. $12u^2 + 11u$

   c. $12u^2 + 5u$

   d. $12u^2 + 5$

   e. None of the above

25. Simplify: $5x - 3\{2x - 2[x - 2(1-x)]\}$

   a. $17x - 12$

   b. $17x + 12$

   c. $-17x - 12$

   d. $12x - 17$

   e. None of the above

26. Which of the following polynomials is equal to $\left(\dfrac{1}{2}x - 5\right)^2$?

   a. $\dfrac{1}{4}x^2 + 5x - 25$

   b. $\dfrac{1}{4}x^2 - 5x + 25$

   c. $\dfrac{1}{4}x^2 + \dfrac{5}{2}x - 25$

   d. $\dfrac{1}{4}x^2 - \dfrac{5}{2}x + 25$

   e. None of the above

27. Which of the following polynomials is equal to the product of $x - 1$ and $2x^2 - x - 1$?

   a. $x^3 + x^2 + x + 1$

   b. $2x^3 - 3x^2 + 2x - 1$

   c. $x^3 - x^2 - 1$

d. $2x^3 - 3x^2 + 1$

e. None of the above

28. Which of the following shapes has the greatest area?

    a. A circle with radius 4

    b. A rectangle with width 3 and height 5

    c. A square with length 4

    d. A right triangle with legs of length 6 and 8

    e. A right triangle with a hypotenuse of length 10 and one leg of length 4

29. Which of the following situations involve traveling at the greatest speed?

    a. Driving 330 miles in 5.5 hours

    b. Driving 550 miles in 10 hours

    c. Driving 145 miles in 2.5 hours

    d. Driving 189 miles in 3 hours

    e. Driving 400 miles in 8 hours

30. Which of the following has the greatest perimeter?

    a. A square with side length 5

    b. A rectangle with length 6 and height 2

    c. A right triangle with legs having length 12 and 5

    d. A circle of radius 3

    e. A semi-circle of radius 5

Solutions to the 30 multiple choice questions:

1. c	6. b	11. c	16. a	21. c	26. b
2. c	7. b	12. a	17. a	22. d	27. d
3. a	8. a	13. b	18. c	23. a	28. a
4. b	9. d	14. c	19. d	24. b	29. d
5. c	10. b	15. b	20. b	25. a	30. c

Index

DENISE SZECSEI earned Bachelor of Science degrees in physics, chemistry, and mathematics from the University of Redlands, and she was greatly influenced by the educational environment cultivated through the Johnston Center for Integrative Studies. After graduating from the University of Redlands, she served as a technical instructor in the U.S. Navy. After completing her military service, she earned a PhD in mathematics from the Florida State University. She recently returned to graduate school to study epidemiology and biostatistics at the University of Iowa. She has been teaching since 1985, and hopes that the FSU Seminoles and the UI Hawkeyes never meet in a BCS bowl game.